Visual Basic Quickstart Guide

Improve your programming skills and design applications that range from basic utilities to complex software

Aspen Olmsted

BIRMINGHAM—MUMBAI

Visual Basic Quick Start Guide

Group Product Manager: Kunal Sawant

Publishing Product Manager: Akash Sharma

Senior Editor: Rounak Kulkarni

Technical Editor: Shruti Thingalaya

Copy Editor: Safis Editing

Project Coordinator: Deeksha Thakkar

Proofreader: Safis Editing

Indexer: Tejal Daruwale Soni

Production Designer: Joshua Misquitta

DevRel Marketing Coordinator: Sonia Chauhan

Business Development Executive: Samriddhi Murarka

First published: October 2023

Production reference: 1210923

Published by Packt Publishing Ltd

Grosvenor House

11 St Paul's Square

Birmingham

B3 1RB

ISBN 978-1-80512-531-0

www.packtpub.com

To the surviving family members of the many computer science students who were taken too soon. Their creativity was too much for this world and I am sure they are developing great software projects in the next world.

– Aspen Olmsted

Contributors

About the author

Aspen Olmsted is an associate professor and program director at Wentworth Institute of Technology in the Computer Science department. He obtained a Ph.D. in Computer Science and Engineering from the University of South Carolina. Before his academic career, he was CEO of Alliance Software Corporation. Alliance Software developed N-Tier enterprise applications for the performing arts and humanities market. Dr. Olmsted's research focuses on developing algorithms and architectures for distributed enterprise solutions that guarantee security and correctness while maintaining high availability. Aspen mentors over a dozen graduate and undergraduate students in his Secure Data Engineering lab each year.

I would like to, first and foremost, thank my loving and patient wife, Kirsten, and children, Freya and Seamus, for their continued support, patience, and encouragement throughout the long process of writing this book.

About the reviewer

Muhammad Hameem Ibne Kabir is passionate about visual coding and secure design and is presently working as a freelance developer. He studied cybersecurity at New York University. He has also done a number of courses on computer science from MIT and Harvard. Muhammed has designed user-friendly and phonetically interrelated keyboard layouts for different languages. He has also developed web-based and desktop apps to run them.

Table of Contents

2

Console Input and Output   17

3

Data Types and Variables   29

4

Decision Branching   39

5

Iteration   47

6

Functions and Procedures   53

7

Project Part I   59

Part 2: Visual Basic Files and Data Structures

8

9

10

Collections   95

11

Project Part II   103

Part 3: Object-Oriented Visual Basic

12

Object-Oriented Programming   125

Part 4: Server-Side Development

17

18

19

20

Preface

In this book, you will learn about the Visual Basic language and all its different use cases. This journey will enable you to apply that understanding to maintain, enhance, administer, and secure Visual Basic console or Windows applications, Visual Basic websites, or Visual Basic scripts. This book starts from an understanding that many programmers will need to use Visual Basic for many different purposes. In each chapter, we highlight topics across the many different use cases Visual Basic is used for.

When learning how to program in a new language, you always want to give yourself time to digest the concepts. Many iterations of the topic help to solidify it in your mind. To get the most out of this book, you are encouraged to have your computer open as you read the book. Another terrific approach is to read a chapter for the first time and then go to your computer and revisit the examples by plugging them in yourself on your computer.

I also encourage you to dig deep into your creativity and think of ways to use the skills we cover in a creative project. I suggest several projects in the last chapter of the book, but you should try to find something meaningful for you. Good luck!

Who this book is for

If you work in software engineering, information technology, or cybersecurity and need to maintain, enhance, administer, or secure Visual Basic console or Windows applications, Visual Basic websites, or Visual Basic scripts, then this book is for you.

At the time of writing this book, Microsoft is reducing its investment in Visual Basic. Despite that, there are many software development jobs that involve maintaining, enhancing, administering, and defending Visual Basic programs, websites, and scripts. These jobs will continue to exist for decades to come. This book is designed for software developers and web developers who find themselves in one of those jobs or want to acquire one of those jobs. This book presents all the different use cases of Visual Basic in a concise way.

What this book covers

Chapter 1, The Visual Basic Family of Programming Languages, provides an overview of VB and its use cases over the years.

Chapter 2, Console Input and Output, teaches you how to retrieve keyboard input and send the results of computations to the computer screen.

Chapter 3, Data Types and Variables, discusses the different ways to store data in the computer memory during the program execution.

Chapter 4, Decision Branching, talks about how to build different pathways inside of our code based on an external environment such as input from the user.

Chapter 5, Iteration, shows how to repeat sections of code either a fixed number of times or until an environment is in a certain state.

Chapter 6, Functions and Procedures, dwells on modularizing sections of code for reuse. This makes it easier to have fewer bugs by only having one copy of code that does a specific thing.

Chapter 7, Project Part I, enables you to apply the programming skills you have learned to this point.

Chapter 8, Formatting and Modifying Data, explores the concept of rearranging data to use for display or for saving externally.

Chapter 9, File Input and Output, shows how to read and write data from and to files on the disk drive.

Chapter 10, Collections, discusses storing multiple values in one variable and then accessing those variables in code.

Chapter 11, Project Part II, provides an opportunity to apply the programming skills we have learned up to this point.

Chapter 12, Object-Oriented Programming, deals with arranging code into objects. Each object can have a state and actions it can perform.

Chapter 13, Inheritance, teaches us how to reuse code through the creation of taxonomies where we specify generic functionality higher in the tree.

Chapter 14, Polymorphism, shows us how to allow VB to execute the proper version of a method at runtime based on the context of the running program.

Chapter 15, Interfaces, explores code reuse through the definition of contracts for implementations that can be developed later but executed by our code.

Chapter 16, Project Part III, allows us to apply the programming skills we have learned up to this point.

Chapter 17, The Request and Response Model, discusses the web application communication model and how we send messages between partitions of our application.

Chapter 18, Variable Scope and Concurrency, explores handling multiple users on a web application.

Chapter 19, Project Part IV, provides a final opportunity to try out what we have learned.

Chapter 20, Conclusions, reviews the topics covered and suggests further learning and projects.

To get the most out of this book

We cover many different platforms and use cases of Visual Basic in this book. You should focus on the VB family members that make sense to you at the time of reading and revisit the book later to think about the other family members. No prior programming experience is required, but if you have programmed in another language, you will find VB comes easily to you.

Software/hardware covered in the book	Operating system requirements
VB.NET	Windows, macOS, or Linux
ASP.NET	Windows
VB6	Windows
VBA	Windows
VBScript	Windows

For VB6, we recommend you install Windows XP in a virtual machine such as VirtualBox.

If you are using the digital version of this book, we advise you to type the code yourself or access the code from the book's GitHub repository (a link is available in the next section). Doing so will help you avoid any potential errors related to the copying and pasting of code.

Download the example code files

You can download the example code files for this book from GitHub at `https://github.com/PacktPublishing/Learn-Visual-Basics-Quick-Start-Guide-`. If there's an update to the code, it will be updated in the GitHub repository.

We also have other code bundles from our rich catalog of books and videos available at `https://github.com/PacktPublishing/`. Check them out!

Conventions used

There are a number of text conventions used throughout this book.

`Code in text`: Indicates code words in text, database table names, folder names, filenames, file extensions, pathnames, dummy URLs, user input, and Twitter handles. Here is an example: "The following is sample Visual Basic.NET Windows Form code that would be stored in a file, with a name such as `program.vb`."

A block of code is set as follows:

```
Public Class Form1
    Private Sub Form1_Load(sender As Object, e As EventArgs) Handles
MyBase.Load
        MsgBox("Hello, World!")
    End Sub
End Class
```

Bold: Indicates a new term, an important word, or words that you see onscreen. For instance, words in menus or dialog boxes appear in **bold**. Here is an example: "Choose **Standard EXE** as the project type."

> **Tips or important notes**
> Appear like this.

Get in touch

Feedback from our readers is always welcome.

General feedback: If you have questions about any aspect of this book, email us at customercare@ packtpub.com and mention the book title in the subject of your message.

Errata: Although we have taken every care to ensure the accuracy of our content, mistakes do happen. If you have found a mistake in this book, we would be grateful if you would report this to us. Please visit www.packtpub.com/support/errata and fill in the form.

Piracy: If you come across any illegal copies of our works in any form on the internet, we would be grateful if you would provide us with the location address or website name. Please contact us at copyright@packt.com with a link to the material.

If you are interested in becoming an author: If there is a topic that you have expertise in and you are interested in either writing or contributing to a book, please visit authors.packtpub.com.

Share your thoughts

Once you've read *Visual Basic Quickstart Guide*, we'd love to hear your thoughts! Scan the QR code below to go straight to the Amazon review page for this book and share your feedback.

https://packt.link/r/1805125311

Your review is important to us and the tech community and will help us make sure we're delivering excellent quality content.

Download a free PDF copy of this book

Thanks for purchasing this book!

Do you like to read on the go but are unable to carry your print books everywhere?

Is your eBook purchase not compatible with the device of your choice?

Don't worry, now with every Packt book you get a DRM-free PDF version of that book at no cost.

Read anywhere, any place, on any device. Search, copy, and paste code from your favorite technical books directly into your application.

The perks don't stop there, you can get exclusive access to discounts, newsletters, and great free content in your inbox daily

Follow these simple steps to get the benefits:

1. Scan the QR code or visit the link below

https://packt.link/free-ebook/9781805125310

2. Submit your proof of purchase

3. That's it! We'll send your free PDF and other benefits to your email directly

Part 1:
Visual Basic
Programming and Scripting

In the first part of this book, we will focus on teaching you the basic programming skills needed to write core code in any of the Visual Basic family members. If you have learned any other programming language, all of these topics will be familiar, as most programming languages have similar topics.

This part has the following chapters:

- *Chapter 1, The Visual Basic Family of Programming Languages*
- *Chapter 2, Console Input and Output*
- *Chapter 3, Data Types and Variables*
- *Chapter 4, Decision Branching*
- *Chapter 5, Iteration*
- *Chapter 6, Functions and Procedures*
- *Chapter 7, Project Part I*

1

The Visual Basic Family of Programming Languages

People have been programming Visual Basic for over 23 years. Microsoft has deployed Visual Basic in many different environments over the years, including Windows Desktop development with Classic Visual Basic and Visual Basic.NET Windows Forms. Visual Basic has also been used as a server-side programming language in Classic ASP and ASP.NET. Visual Basic for Applications has been used to embed Visual Basic in both Microsoft Office products along with many third-party products. Lastly, many scripts have been used to control specialized hardware or automate administrative tasks. You will walk away from this chapter with a strong foundation to meet your Visual Basic requirements.

In this chapter, we're going to cover the following main topics:

- Visual Basic.NET Windows Forms
- Visual Basic Classic
- **Visual Basic for Applications** (**VBA**)
- **Visual Basic Scripting Edition** (**VBScript**)
- Classic ASP
- ASP.NET

Technical requirements

Each Visual Basic family member will have different technical requirements. We will provide installation options and steps in each subsection for the different family members. Some of these products are old and should be installed in a **virtual machine** (**VM**) or development machine that does not need to maintain security. All example code for this book is available at the following GitHub repository: `https://github.com/PacktPublishing/Learn-Visual-Basics-Quick-Start-Guide-`.

Programming with Visual Basic.NET Windows Forms

Visual Basic.NET Windows Forms is a **user interface (UI)** framework to create desktop applications on the Windows operating system. It allows developers to create visually appealing and interactive applications using a drag-and-drop interface, without needing to write all the code from scratch.

Windows Forms provides a set of pre-built controls such as buttons, textboxes, labels, and menus that can be easily placed on a form to create a UI. These controls can be customized by changing properties such as color, size, and font, and they can be programmed to respond to user input and events.

Visual Basic.NET is a programming language that can be used to write code for Windows Forms applications. It is a high-level language that is easy to learn and use, and it provides a wide range of features to develop complex applications.

Some of the features of Visual Basic.NET Windows Forms include the following:

- **Rapid application development**: Windows Forms provides a quick and easy way to create desktop applications using a visual designer.

- **Data binding**: Windows Forms supports data binding, which allows developers to connect controls to data sources such as databases or XML files.

- **Advanced controls**: Windows Forms includes a set of advanced controls such as `DataGridView` and the Chart control, which can be used to create more sophisticated UIs.

- **Object-oriented programming (OOP)**: Visual Basic.NET supports OOP, which allows developers to write reusable and maintainable code.

OOP

OOP is a programming paradigm based on the concept of objects, which are entities that contain both data and behavior. In OOP, objects are created from classes, which define the data and behavior the objects will have. The main principles of OOP include the following:

- **Encapsulation**: This principle involves bundling data and behavior into objects and hiding the implementation details from the outside world. Encapsulation helps to ensure that objects are used correctly and that the data they contain is not modified in unintended ways.

- **Inheritance**: This principle involves creating new classes by extending existing classes. Inheritance allows new classes to inherit the data and behavior of their parent classes, and it also allows them to add their unique features.

- **Polymorphism**: This principle allows objects of different classes to be treated as the same object type. Polymorphism is achieved through interfaces or abstract classes, which define a set of standard methods that can be used by all objects that implement the interface or inherit from the abstract class.

OOP has many advantages, including improved code organization, increased code reusability, and better maintainability.

Overall, Visual Basic.NET Windows Forms is a powerful tool to create desktop applications on the Windows platform, and it is a popular choice among developers due to its ease of use and wide range of features.

Unfortunately, Microsoft has announced they will not be evolving the Visual Basic.NET language in the future. This announcement does not mean the language is dead – on the contrary, billions of lines of production code out there will continue to execute for many decades. It just means new language features will not be added to the language. Now that we have familiarized ourselves with Visual Basic.NET Windows Forms, let us look at coding a simple example.

Hello, World! in Visual Basic.NET Windows Forms

Visual Basic.Net Windows Forms code is stored in text files with an extension of `.vb`. The following is sample Visual Basic.NET Windows Form code that would be stored in a file, with a name such as `program.vb`. Running the program will display a message box that displays `Hello World!` on startup:

```
Public Class Form1
   Private Sub Form1_Load(sender As Object, e As EventArgs) Handles
MyBase.Load
        MsgBox("Hello, World!")
   End Sub
End Class
```

Next, we will see how to implement this code.

Installing Visual Basic.NET

The following steps will install Visual Basic.NET onto a Windows machine:

1. Download your Visual Studio Community Edition from this link: `https://learn.microsoft.com/en-us/visualstudio/install/install-visual-studio?view=vs-2022`.

2. Double-click the `VisualStudioSetup.exe` bootstrapper to start the installation.

3. Choose the **.NET desktop development** workload.

4. Let the installer finish.

Most users will use the Microsoft Windows **operating system (OS)** and Microsoft's implementation of Visual Basic.NET, but there are projects and products that allow installation on different OSs, which we will discuss next.

Alternative installations of Visual Basic.NET

The following list shows other options to work with Visual Basic.NET-compatible code:

- **Mono**: Mono is an open source development platform based on the .NET framework. It includes a Visual Basic.NET compiler and VB.NET runtime. You can download Mono from here: `http://mono-project.com`.

- **.NET SDK**: The .NET **Software Development Kit** (SDK) is a set of tools, libraries, and components to develop applications that target the .NET platform. It includes everything you need to build and deploy .NET applications, including a compiler, runtime, and various libraries.

  The .NET SDK is available for multiple platforms, including Windows, Linux, and macOS. It supports multiple programming languages, including C#, F#, and Visual Basic.

  With the .NET SDK, developers can build a variety of applications, such as desktop applications, web applications, mobile applications, games, and more. It also includes tools to build and manage cloud-based applications, including tools to deploy to Microsoft Azure.

  The .NET SDK is constantly evolving, with new features and updates being added regularly. It is an essential tool for any .NET developer, providing everything they need to create high-quality, robust applications. It can be downloaded from `http://dotnet.microsoft.com`.

Next up, we will program a simple example in VB.NET.

Coding Visual Basic.NET Windows Forms Hello, World!

The following steps will allow you to code and execute a Visual Basic.NET Windows Forms version of `Hello World!`:

1. Start **Microsoft Visual Studio**.
2. Choose **Create New Project**.
3. Choose **Windows Forms App** and click **Next**.
4. Accept **Default Project Name** and **Solution Name**.
5. Change **Location** if you prefer your code in a different folder.
6. Accept the default framework.
7. Click **Create Project**.
8. When the **Designer** loads, double-click on the form.
9. Enter the `MsgBox` line from the code example that we saw earlier.

10. Click the green triangle on the toolbar above your code to run your program.

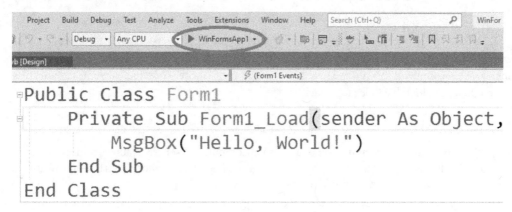

Figure 1.1 – Running your program

We will work with this same application example in different Visual Basic family members as we continue our journey.

Programming with Visual Basic Classic

Visual Basic Classic (also known as **VB6**) is a programming language and **integrated development environment** (**IDE**) that was developed by Microsoft. It was released in 1998 and was widely used to develop desktop applications for the Windows operating system.

VB6 was the last version of the Visual Basic programming language before the introduction of .NET Framework, and it uses a syntax similar to earlier versions of Visual Basic. The language was designed to be easy to learn and use, and it was popular among beginners and experienced developers alike.

VB6 applications can be developed using the Visual Basic IDE, which includes a visual form designer, code editor, and debugger. The language supports OOP concepts, such as encapsulation, inheritance, and polymorphism.

Although Microsoft has stopped supporting VB6, many legacy applications continue to use the language, and it remains a popular choice for small-scale Windows desktop application development.

Hello, World! in Visual Basic Classic

Visual Basic Classic code is stored in text files with an extension of `.bas`:

```
Private Sub Form_Load()
    MsgBox "Hello, World!"
End Sub
```

You will implement this code next.

Installing Visual Basic 6

The following steps are designed to install the original Visual Basic 6 Enterprise onto a Windows machine. We recommend this is done in a virtual machine or Docker container, as some support files are very old and vulnerable to legacy attacks.

> **VM**
>
> A VM is a software-based emulation of a physical computer or server. It allows multiple OSs to run simultaneously on a single physical machine. Each OS can access virtual hardware resources such as CPU, RAM, and storage.
>
> VMs are commonly used in various contexts, including software development, testing, and deployment, hosting multiple servers on a single physical machine, and running legacy applications that require specific OSs.
>
> There are several popular VM software packages, including VMware, VirtualBox, Hyper-V, and KVM, each with advantages and disadvantages.

1. Download your language/architecture version of `cd image` for Visual Basic 6.0 Enterprise from **WinWorldPC**: `https://winworldpc.com/product/microsoft-visual-bas/60`.

2. The downloaded file is a compressed archive in the `7zip` format. You can extract it with a ZIP utility similar to `http://www.7-zip.org`.

3. Either mount the ISO image as a virtual drive or burn it to a CD.

4. Run the `setup.exe` program to install Visual Studio. (The key to enter is below the product links on the WinWorldPC website.)

5. The setup will try to install an old Java VM. You can either allow it to do so, or you can refer to the many blogs online that describe how to trick the setup program into not trying to install the VM.

6. The setup program will ask you to reboot if you install the Java VM in the previous step. If the program does not continue automatically, just go back to *step 3* and start again, and it will continue from where it left off.

7. Choose **Product Install**.

8. Choose **Visual Basic Enterprise**.

Again, most users will want to use the Microsoft version of Visual Basic Classic. There are alternatives, which we will explore next. We will start by looking at some alternatives to Microsoft Visual Basic 6 that try to be compatible with the VB6 syntax.

Alternative installations

The following list shows other options to work with Visual Basic 6-compatible code:

- **WINE**: This is an application that allows you to run Windows programs on Linux systems. At the time of writing, Visual Basic 6 is listed at *silver*-level compatibility. Silver level typically means there will be some visual issues, but the program will still run.

- **TwinBasic**: TwinBasic is a modern BASIC programming language. The language aims for 100% backward compatibility with existing VB6/VBA projects. In addition, TwinBasic aims to become a complete replacement development environment, offering considerable new features and improvements over the VB6 IDE. It can be downloaded from here: `https://twinbasic.com/`.

- **B4X**: The B4X suite includes **Basic for Android (B4A)**, **Basic for iOS (B4i)**, **Basic for Java (B4J)**, and **Basic for Arduino (B4R)**. The syntax is very much VB-inspired, but it is not 100% compatible with VB6 syntax. Download it from here: `https://www.b4x.com/`.

- **RAD Basic**: The RAD Basic project goal is to have a 100% VB6-compatible development environment. You can download it from `https://www.radbasic.dev/`.

- **VB6 Installer**: This tool makes the native installation of the original Visual Basic 6 smoother on Windows 10 and 11. Download it from `http://nuke.vbcorner.net/VS6Installer/tabid/125/language/en-US/Default.aspx`.

Next up, we will examine how to plug the example code into Visual Basic 6 and run the program.

Running the Visual Basic 6 Hello, World! program

The following steps will allow you to code and execute a Visual Basic 6 version of `Hello World!`:

1. Start **Microsoft Visual Basic 6**.
2. Choose **Standard EXE** as the project type.
3. When the designer loads, double-click inside the rectangle representing your form.
4. Enter the `MsgBox` line from the code example seen earlier.
5. Click the green triangle on the toolbar above your code to run your program.

Let's move on to look at VBA.

Programming with VBA

VBA is a programming language that automates tasks in Microsoft Office applications, such as Excel, Word, PowerPoint, and Access. VBA is primarily used to automate repetitive tasks and interact with external applications.

VBA is a powerful language that can import and translate data, automate sequences of calculations, and create custom forms and specialized reports. These powers lead VBA to be used by finance professionals and data analysts who work with large amounts of data and perform complex analyses. VBA is a powerful language that allows users to create functions, perform error handling, and use advanced programming techniques such as loops and conditional statements. VBA is also licensed from Microsoft or reproduced by third-party companies. A small sample of applications that host VBA includes **WordPerfect**, **LibreOffice**, and **CorelDRAW**.

Hello, World! in VBA

VBA code is stored in text files with an extension of `.bas` when exported from the host. Often, the code is stored in the host document, depending on the application host. The following is sample VBA macro code, with a macro named `test`. Running the macro will display a message box that displays `Hello, World!` on startup:

```
Sub test()
    MsgBox ("Hello, World!")
End Sub
```

You will implement this code shortly.

> **Installing VBA**
>
> VBA is installed by default with Microsoft Office applications. Other hosts may require you to choose the option in the host installation.

Coding a VBA Hello, World! from Excel

The following steps will allow you to code and execute a VBA macro that displays `Hello World!`:

1. Start Microsoft Excel.
2. Click on the **View** menu.
3. Click the **Macros** button, and choose **View Macros** from the drop-down menu.
4. Type the name `test` as the macro name.
5. Click the **Create** button.
6. When the designer loads, double-click inside the rectangle representing your form.
7. Enter the `MsgBox` line from the code example seen earlier.
8. Click the green triangle on the toolbar above your code to run your macro.

Up next, we will look at the version of Visual Basic used for scripting.

Programming with VBScript

VBScript is a scripting language that is based on the Visual Basic programming language. It is primarily used to create simple scripts that automate tasks in Windows environments.

VBScript is a lightweight language that is easy to learn and use. It is commonly used to automate repetitive tasks, such as file manipulation, data processing, and system administration. It can be used to interact with Windows components, such as the registry, **Windows Management Instrumentation (WMI)**, and Active Directory.

VBScript is an interpreted language, which means that it is executed directly by the OS without the need for a compiler. It can be run from the command line or embedded within other applications, such as Microsoft Excel or Internet Explorer.

VBScript is supported by all modern versions of Windows and is often used in combination with other scripting languages, such as PowerShell and Batch files.

Hello, World! in VBScript

VBScript code is stored in text files with an extension of `.vbs`. Often, the code is stored in the filesystem where the script will be executed. The following is sample VBScript code that is stored in a file named `test.vbs`. Running the script will display a message box that displays `Hello, World!` on startup:

```
MsgBox "Hello, World!"
```

You will implement this code shortly.

> **Installing VBScript**
>
> In modern Windows versions, VBScript is part of Windows Script Host. By default, the primary scripting language with a new Windows installation is VBScript.

Coding a Hello, World! VBScript with Notepad

The following steps will allow you to code and execute a VBScript that displays `Hello World`:

1. Start Notepad.
2. Enter the `MsgBox` line from the code example shown earlier.
3. Save the file with the name `test.vbs` in your root folder.
4. Run `CMD` to go to Command Prompt.
5. Type `cscript test.vbs` and press the *Enter* key.

In the next section, VBScript will be used on the server side of web development.

Programming with Classic ASP

Classic **Active Server Pages (ASP)** is a server-side scripting technology used to create dynamic web pages and applications. Microsoft's first version of ASP was in the late 1990s as part of the Microsoft ActiveX technology.

Classic ASP uses VBScript as its default scripting language, but it can also be used with other scripting languages. Some examples of other languages include **JScript** and **PerlScript**. VBScript allows developers to program scripts that run on the server side into HTML pages. This scripting allows for the creation of dynamic content that changes based on environmental variables, works with data in databases, and performs other server-side tasks.

Classic ASP historically was used to develop web applications that ran on **Microsoft Internet Information Services (IIS)** servers. Classic ASP has often been used with data stored in **Microsoft SQL Server**, **MySQL**, or **Oracle** databases.

Despite being replaced by ASP.NET, Classic ASP is still widely used in some organizations and industries, particularly to maintain legacy applications and support older systems.

Hello, World! in Classic ASP

Classic ASP code is stored in text files with an extension of .asp. The code is stored in the filesystem where **Internet Information Services (IIS)** hosts websites. The following is sample Classic ASP code that is stored in a file named test.asp. Displaying the web page will display a message in HTML with the words Hello, World!:

```
<!DOCTYPE html>
<html>
<body>
<%
  Response.Write("Hello World!")
%>
</body>
</html>
```

You will implement this code shortly. The code is a mix of VBScript, with the <% %> tags and HTML outside the tags. IIS will replace the code in the <% %> tags with the output from the execution.

Installing Classic ASP on Windows

Classic ASP is installed with IIS. Follow these steps to install both IIS and the ASP development features on your development box or inside a VM:

1. Click **Start Menu** in Windows.
2. Click on **Control Panel**.

3.   In **Control Panel**, click **Programs and Features**.

4.   Click **Turn Windows features on or off**.

5.   Click on the + symbol to expand **Internet Information Services**.

6.   Expand **World Wide Web Services**.

7.   Expand **Application Development Features**.

8.   Select **ASP**.

9.   Click **OK**.

Classic ASP was so popular that alternative products were developed to be compatible with it, which we will look at next.

Alternative installations

ASP Classic is not compiled by default, as it is an interpreted language. The server reads the ASP Classic code, interprets it, and generates HTML output that is sent to the client's web browser. This interpretation process can result in slower performance compared to compiled languages, such as C# or Java.

However, there are third-party tools available that can compile ASP Classic code into machine code, which can improve performance and provide other benefits, such as increased security and the ability to obfuscate code. Some popular tools to compile ASP Classic code include the following:

- **ASP Compiler**: A commercial tool that compiles ASP Classic code into native machine code. It also provides a number of features such as code optimization and obfuscation.

- **SmartAssembly**: A commercial tool that can compile ASP Classic code into a native Windows executable. It also provides features such as code optimization, obfuscation, and error reporting.

- **CodeWall**: A free tool that can compile ASP Classic code into a native Windows executable. It provides features such as code obfuscation and resource compression.

It's worth noting that while these tools can provide performance benefits, they can also introduce additional complexity and require changes to the deployment process. Therefore, it's important to carefully evaluate the benefits and drawbacks of compiling ASP Classic code before deciding to use one of these tools.

Coding Classic ASP Hello, World! with Notepad

The following steps will allow you to code and execute a VBScript that displays `Hello World!`:

1.   Start Notepad.

2.   Enter all the code from the code example shown earlier.

3. Save a file with the name `test.asp` in the root web directory (this is typically `c:\inetpub\wwwroot`).

4. Open a web browser.

5. Enter the URL: `http://127.0.0.1/test.asp`.

Next, we will look at the newest version of server-side code from Microsoft.

Programming in ASP.NET

ASP.NET is a web application framework developed by Microsoft that allows developers to create dynamic web applications using a variety of programming languages, such as C# and Visual Basic.NET. It provides several features and tools that make it easier to build, test, and deploy web applications.

ASP.NET allows developers to build web applications using different programming patterns, including **Model-View-Controller** (**MVC**) and Web Forms architecture. The MVC pattern provides a definite separation of concerns between the three main components in the architecture – model, view, and controller. The model represents the data and business logic, the view is used to display the data to the user, and the controller handles user input and interacts with the model and the view. In contrast, Web Forms provides more of a desktop development approach, allowing developers to create web pages containing server-side controls and events.

ASP.NET provides several features that make it easier to build robust and scalable web applications. These features include built-in support for caching, session management, and authentication. ASP.NET also integrates with other Microsoft technologies such as **SQL Server**, **Windows Communication Foundation** (**WCF**), and **Windows Workflow Foundation** (**WF**).

ASP.NET is compatible with a wide range of web servers and operating systems, including Microsoft IIS, Apache, and nginx. ASP.NET also supports a variety of client-side scripting technologies, such as JavaScript and Ajax. These client-side technologies allow a developer to create interactive and responsive web applications.

Hello, World! in ASP.NET

ASP.NET code is stored in text files with an extension of `.vb`. The file is code behind and has the same starting name as the HTML stored in the ASPX file. The code is stored in the filesystem where IIS hosts websites. The following is sample ASP.NET code that is stored in a file named `WebForm1.aspx.vb`. Displaying the web page will display a message in HTML with the words `Hello, World!`:

```
Protected Sub Page_Load(ByVal sender As Object, ByVal e As System.EventArgs)
    Handles Me.Load
        Response.Write("Hello, World!")
    End Sub
```

You will implement this code shortly.

Installing ASP.NET

ASP.NET Development is installed with Visual Studio Community Edition. Follow these steps to install the ASP.NET development features:

1. Download Visual Studio Community Edition 2019 (Visual Basic is not supported in ASP.NET apps in the 2022 edition) from `https://visualstudio.microsoft.com/vs/older-downloads/`.

2. Double-click the `VisualStudioSetup.exe` bootstrapper to start the installation.

3. Choose the `ASP.NET and web development` workload.

4. Let the installer finish.

ASP.NET is designed to be installed on a production server without the development tools. We will look at that installation next.

Alternative installations

Windows Server – ASP.NET is installed with IIS in Windows Server production environments. Follow these steps to install both IIS and the ASP.NET development features in your production box:

1. Click **Start Menu** in Windows.

2. Click **Control Panel**.

3. In **Control Panel**, click **Programs and Features**.

4. Click **Turn Windows Features on or off**.

5. Expand and navigate to **Internet Information Services** | **World Wide Web Services** | **Application Development Features**.

6. Select the latest ASP.NET version.

7. Click **OK**.

> **Note**
>
> Microsoft maintains instructions on installing ASP.NET on Linux servers: `https://learn.microsoft.com/en-us/dotnet/core/install/linux`.

Coding ASP.NET Hello, World! with Visual Studio

The following steps will allow you to code and execute an ASP.NET web page that displays `Hello World!`:

1. Start Microsoft Visual Studio.
2. Choose **Create New Project**.
3. Choose **ASP.NET Web Application** and click **Next**.
4. Accept **Default Project Name** and **Solution Name**.
5. Change the location if you prefer your code in a different folder.
6. Accept the default framework.
7. Click the **Create** button.
8. Choose **Web Forms**.
9. When the designer loads, navigate in the solution explorer and right-click on the solution.
10. Choose **Add** from the pop-up menu.
11. Choose **New Item** from the pop-up menu.
12. Choose **Web Form Visual Basic**.
13. Click the **Add** button.
14. Right-click on the HTML editor, and choose **View Code** from the pop-up menu or **Source** from the tabs.
15. Enter the `Response.write` line from the earlier code example.
16. Click the green triangle on the toolbar above your code to run your program.

That concludes our high-level investigation into the different family members of Visual Basic.

Summary

In this chapter, you were introduced to six different environments where Visual Basic programming is relevant. Over the past three decades, programmers have used all these methods to solve business needs with Visual Basic. In future chapters, we will build up your understanding of Visual Basic syntax and semantics and apply that knowledge to these six different environments. In *Chapter 2*, we will specifically look at how we can get data from users as input and send information back to them as output.

2

Console Input and Output

Every algorithm in computing requires taking in some data as input and outputting the results somehow. In this chapter, we will discuss simple ways to interact with a user in each Visual Basic family member. For gathering input from the user in this chapter, we will focus on reading data from a keyboard used by the user on the client of the applications we will develop. For output, we will send raw data back to the user's screen in a simple model of rows of text with line delimiters.

In this chapter, we're going to cover the following main topics:

- Programming input and output with Visual Basic.NET
- Programming input and output with Visual Basic Classic
- Programming input and output with **Visual Basic for Applications (VBA)**
- Programming input and output with VBScript
- Programming input and output with Classic ASP
- Programming input and output with ASP.NET

Technical requirements

Please make sure you complete the steps to install and code with the Visual Basic family member from *Chapter 1*.

Programming input and output with Visual Basic.NET

The `Console` class in Visual Basic.NET allows a program to interact with a raw console for input and output. In the previous chapter, we utilized a project template for creating the Windows Forms app. This chapter will use the Console App project template to give us easy access to the `Console` class. Console App projects can run on Microsoft Windows, Linux, and macOS operating systems.

For output, we can utilize the `Write` and `WriteLine` methods of the `Console` class to output data to the screen. Here are some details to think about:

- The `Console.Write` and `Console.WriteLine` methods take a string parameter and send it to the console

- The `Console.Write` method leaves the cursor at the end of the output so further output continues on the same line

- `Console.WriteLine` sends the cursor to the following line

We can use the `ReadLine` method of the `Console` class to input data from the keyboard. Here are some details to think about:

- `Console.ReadLine` returns a string of text that was read up to the enter key

- We can use the string returned from the method in any expression

Next, let's look at a simple interaction between the code and the user.

Simple interaction in Visual Basic.NET

The following Visual Basic.Net code will utilize the `Console.WriteLine` method to request the user to give their name. The code uses `Console.ReadLine` to retrieve the user's name. Lastly, the `Console.WriteLine` displays the greeting to the console.

```
Module Program
    Sub Main(args As String())
        Dim name As String
        Console.WriteLine("What is your name?")
        name = Console.ReadLine()
        Console.WriteLine("It is nice to meet you " + name)
    End Sub
End Module
```

The first line in the code declares a variable named Name to store the value returned from the keyboard. The second line in the code will display the message to the user and a question about their name. The third line will wait for the user to enter a value followed by the enter key. Finally, the fourth line will combine the literal string, `"It is nice to meet you"` with the value stored in the Name variable and displayed on the screen. You will implement this code next.

Coding a simple interaction in Visual Basic.NET

The following steps allow you to code and execute a Visual Basic.NET Windows Forms version of Hello World:

1. Start Microsoft Visual Studio.
2. Choose to **Create New Project**.
3. Choose **Console (.NET Framework)** and click **Next**.
4. Accept **Default Project Name** and **Solution Name**.
5. Change the location if you prefer your code in a different folder.
6. Accept the default framework.
7. Click **Create Project**.
8. Enter the code example from the earlier section.
9. To run your program, click the green triangle on the toolbar above your code.

As we progress through this chapter, you will see that each VB family member has similar methods for input and output.

Programming input and output with Visual Basic Classic

VB Classic does not have a method to develop console applications. Visual Basic Classic was designed to build applications that run only on Windows. For simple input and output, we can use two functions:

*   `InputBox`: Displays a string passed into the method, draws a box for the user to enter their response to the prompt, and returns the value entered
*   `MsgBox`: Displays a message box with the string passed in

Simple interaction in Visual Basic Classic

Here is the code for a simple interaction between the user and the program in Visual Basic Classic:

```
Private Sub Form_Load()
  Dim name As String
  name = InputBox("What is your name?")
  MsgBox ("It is nice to meet you " + name)
End Sub
```

The first line in the code declares a variable named Name to store the value returned from the keyboard. The second line utilizes the InputBox method to draw an input box for the user to enter their name. The returned value is stored in the variable named Name. The third line will combine the literal string "It is nice to meet you" with the value stored in the name variable and display it in a message box. You will implement this code next.

Coding a simple interaction in Visual Basic 6

The following steps allow you to code and execute a Visual Basic 6 version of Hello World:

1. Start Microsoft Visual Basic 6.
2. Choose **Standard EXE** as the project type.
3. When the Designer loads, double-click inside the rectangle representing your form.
4. Enter the code from the example in the previous section.
5. To run your program, click the green triangle on the toolbar above your code.

In our next section, we will tackle input and output in VBA, which looks very similar to what we did in VB 6.

Programming input and output with VBA

VBA, like Visual Basic Classic, will only run in Windows applications. For simple input and output, we can use two functions:

- InputBox: Displays a string passed into the method, draws a box for the user to enter their response to the prompt, and returns the value entered

- MsgBox: Displays a message box with the string passed in

Simple interaction in VBA

Running the following macro will display an input box that prompts the user for their name and then displays a message box greeting the users:

```
Sub test()
  Name = InputBox("What is your name?")
  MsgBox ("It is nice to meet you " & Name)
End Sub
```

The first line in the code utilizes the InputBox method to draw an input box for the user to enter their name. The returned value is stored in the variable named Name. The second line will combine the literal string "It is nice to meet you" with the value stored in the name variable and display it in a message box. You will implement this code in the following section.

Coding a simple interaction with VBA in Excel

The following steps allow you to code and execute a VBA macro displaying `Hello World`:

1. Start Microsoft Excel.

2. Navigate to the **View** menu.

3. Click the **Macros** button and choose **View Macros** from the drop-down menu.

4. Type the name test into the macro name.

5. Click the **Create** button.

6. When the Designer loads, double-click inside the rectangle representing your form.

7. Enter the two lines from the code example in the previous section.

8. Click the green triangle on the toolbar above your code to run your macro.

We will now move on to input and output in VBScript, which is similar and different from what you have seen so far.

Programming input and output with Visual Basic Script

CScript and **WScript** are two scripting engines used in the Windows operating system to run scripts written in VBScript and other scripting languages.

CScript is a command-line-based scripting engine that runs scripts in a command prompt window. It is primarily designed for running scripts that require no user interface or graphical output and is often used for system administration tasks. CScript is typically used for running scripts in batch files or scheduled tasks, as it provides more control over the execution of the script.

On the other hand, WScript is a GUI-based scripting engine that can display graphical output such as message boxes and dialog boxes. It is primarily designed for running scripts that require user interaction or display information to the user. WScript is typically used for running scripts from within a graphical environment, such as the Windows desktop or Windows Explorer.

CScript and WScript are part of the **Windows Script Host (WSH)** environment and can run scripts written in VBScript, JScript, and other scripting languages. The choice between CScript and WScript depends on the script's requirements and the environment in which it will be run. For this chapter, we will utilize CScript.

Confusingly, both CScript and WScript include a built-in object named WScript. The WScript object contains two beneficial properties for console input and output. The properties we will utilize are the following:

- StdOut: This property is an object that represents the console's Standard Output. A method on this object named Write takes text as a parameter and sends the text to StdOut. By default, StdOut will be the screen but can be redirected on execution to a file or other output device.

- `StdIn`: This property is an object that represents the console's Standard Input. A method on this object named `ReadLine` returns data from `StdIn`. By default, `StdIn` will be the keyboard but can be redirected on execution from a file or other input device.

Simple interaction in VBScript

Here is the code for a simple interaction between the user and the program in VBScript:

```
Dim name
WScript.StdOut.Write("What is your name?")
name = WScript.StdIn.ReadLine()
WScript.StdOut.Write("It is nice to meet you " & name)
```

The first line in the code declares a variable named Name to store the value returned from the keyboard. The second line in the code displays the message to the user with the question about their name. The third line will wait for the user to enter a value followed by the *Enter* key. Finally, the fourth line will combine the literal string `"It is nice to meet you"` with the value stored in the variable named Name and displayed on the screen. You will implement this code in the following section.

Coding a simple interaction in VBScript with Notepad

The following steps allow you to code and execute a VBScript displaying Hello World:

1. Start Notepad.
2. Enter the four lines from the code example from the preceding section.
3. Save the file with the name `test.vbs` in your `root` folder.
4. Run CMD to go to a command prompt.
5. Type `cscript test.vbs` and press the *Enter* key.

We think about input and output from a web application perspective in the following two sections. It may feel very different, but you will get the hang of it quickly.

Programming input and output with Classic ASP

Classic ASP is a server-side scripting language that allows you to generate dynamic content for web pages. With Classic ASP, you can interact with users via **HTML** forms and process the data submitted by users. For example, to interact with users via HTML, you can create HTML forms that collect user information and send that data to a Classic ASP script on the server.

Simple interaction in Classic ASP

To implement the simple interaction example in Classic ASP, we will utilize a single ASP page for simplicity. The HTML form can be stored separately from the ASP page that receives the data. The following is a sample Classic ASP code stored in a single file named test.asp. Displaying the web page will be an input box to enter the user's name. A submit button will be available that calls back to the same page, and if the name is passed in, it will display the greeting back to the user:

```
<!DOCTYPE html>
<html>
<body>
<form method="get">
  What is your name? <input name="name">
  <input type="submit">
</form>
<%
  dim name
  name = Request.QueryString("name")
  if name<>"" then
    Response.Write("It is nice to meet you " & name)
  end if
%>
</body>
</html>
```

You will implement this code in the next section. The code is a mix of VBScript with the <% %> tags and HTML outside the tags. IIS will replace the code in the <% %> tags with the output from the execution. The first line inside the VBScript block declares a variable to hold the name. The second line inside the VBScript block retrieves the name from the request passed to the page. The third line checks if a value of the name was passed in. The fourth line only runs if there is a value for the name passed in, and it will display the greeting back by concatenating the name with the string literal.

Coding a simple interaction in Classic ASP with Notepad

The following steps allow you to code and execute a VBScript displaying Hello World:

1. Start Notepad.
2. Enter all the code from the code example in the previous section.
3. Save the file with the name test.asp in the root web directory (This is typically, c:\ inetpub\wwwroot).
4. Open a web browser.

5. Enter the URL: `http://127.0.0.1/test.asp`.

Figure 2.1 – Simple interaction with Notepad output

The preceding screenshot shows the desired output as visible on the web browser.

In the last section, we will look at a similar approach for input and output used in ASP.NET.

Programming input and output with ASP.NET

ASP.NET utilizes tags to separate the user interface rendered in **HTML**, **JavaScript**, and **CSS** from the Visual Basic that runs on the server. ASP.NET tags are used in the Microsoft ASP.NET web development framework to create dynamic web pages. They are embedded within the **HTML** markup of an ASP. NET page and perform various tasks, such as displaying data from a database, creating user interface elements, and handling user input.

Some common ASP.NET tags include the following:

- `<% %>`: This tag is used to embed server-side code in an ASP.NET page. The code within these tags is executed on the server before the page is sent to the client.

- `<%@ %>`: This tag is used to specify page directives, which are instructions used to configure the behavior of an ASP.NET page. For example, the `<%@ Page %>` directive sets properties such as the page's language, theme, and master page.

- `<asp: %>`: This tag creates ASP.NET server controls and user interface elements that can respond to user events such as clicks or text input. Examples of server controls include buttons, text boxes, and drop-down lists.

- `<%= %>`: This tag outputs dynamic content to an ASP.NET page. Any content written within these tags is evaluated and rendered as HTML when the page is requested.

There are many other ASP.NET tags available, and their use depends on the specific needs of the application being developed.

Simple interaction in ASP.NET

The sample code for ASP.NET code will be stored in two text files. The user interface will be in the ASPX file and contains HTML and ASP.NET tags. The code is stored in the file system where IIS hosts websites. The following is a sample ASP.NET code stored in a file named `WebForm1.aspx`:

```
<%@ Page Language="vb" AutoEventWireup="false" CodeBehind="WebForm1.
aspx.vb" Inherits="WebApplication2.WebForm1" %>

<!DOCTYPE html>

<html xmlns="http://www.w3.org/1999/xhtml">
<head runat="server">
    <title></title>
</head>
<body>
    <form id="form1" runat="server">
        What is your name?
        <asp:TextBox id="name" runat="server"/>
        <asp:Button text="Submit" runat="server"/>
    </form>
</body>
</html>
```

The Visual Basic code is stored in the filesystem where IIS hosts websites. The following is some example VB.NET code stored in a file named `WebForm1.aspx.vb`:

```
Public Class WebForm1
    Inherits System.Web.UI.Page

    Protected Sub Page_Load(ByVal sender As Object, ByVal e As System.
    EventArgs) Handles Me.Load
        Dim name As String
        name = Request.Form("name")
        If Not name Is Nothing Then
            Response.Write("It is nice to meet you " & name)
        End If
    End Sub

End Class
```

The first line inside the VB.NET block inside the `Page_Load` method declares a variable to hold the name. The second line inside the `Page_Load` method retrieves the name from the request passed to the page. The third line in the `Page_Load` method checks if a value of the name was passed in. Finally, the fourth line in the `Page_Load` method only runs if there is a value for the name passed in, and it will display the greeting back by concatenating the name with the string literal.

You will implement this code next.

Coding a simple interaction in ASP.NET with Visual Studio

The following steps will allow you to code and execute an ASP.NET web page that displays Hello World:

1. Start Microsoft Visual Studio 2019.
2. Choose to **Create New Project**.
3. Choose **ASP.NET Web Application** and click **Next**.
4. Accept **Default Project Name** and **Solution Name**.
5. Change the location if you prefer your code in a different folder.
6. Accept the default framework.
7. Click **Create Project**.
8. Choose **Web Forms**.
9. When the Designer loads, navigate to the solution explorer and right-click on the solution.
10. Choose **Add** from the pop-up menu.
11. Choose **New Item** from the pop-up menu.

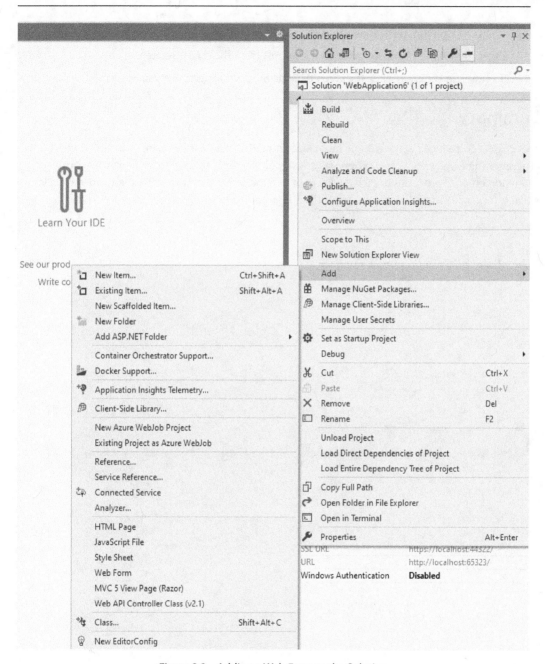

Figure 2.2 – Adding a Web Form to the Solution

12. Choose **Web Form Visual Basic**.

13. Navigate to the source tab and enter the ASPX code from earlier.

14. Right-click on the ASPX source and choose **View Code** from the pop-up menu.

15. Enter the VB.NET code from the code example from the earlier section.

16. To run your program, click the green triangle on the toolbar above your code.

Summary

This chapter introduced you to simple input and output in the six different environments where VB programming is relevant. As we progress in this book, we will build on these tools so you can read and write complete programs in each environment. In the next chapter, we will explore how to declare variables and use them in each member of the VB family.

3

Data Types and Variables

In **Visual Basic** (**VB**), variable declaration defines a variable's name, data type, and initial value (if any) before it is used in the program. Variables are used to store values assigned to them.

In this chapter, we're going to cover the following main topics:

- Choosing data types
- Understanding variable scope
- How you should name variables
- Storing more than one value
- Differences in data types between family members

Choosing data types

Programming languages typically support several data types that can be used to represent different kinds of values. Here are some common data types in programming languages:

- **Integer**: This data type represents whole numbers, both positive and negative
- **Float** or **Double**: This data type represents decimal numbers, also called floating-point numbers
- **Boolean**: This data type represents a logical value, either `true` or `false`
- **Character**: This data type represents a single character or letter, such as *a* or *b*
- **String**: This data type represents a sequence of characters, such as a word or a sentence
- **Array**: This data type represents a collection of elements of the same data type stored in a contiguous memory block
- **Pointer**: This data type represents a memory address, which can be used to access data stored elsewhere in memory

- **Struct** or **Class**: This data type represents a composite data type that can store multiple data types
- **Enumerated**: This data type represents a list of named constants
- **Void**: This data type represents the absence of a value

Different programming languages have different data types available, and some more specific data types depending on their purpose and design. Choosing a data type that is as precise as possible for what you want to store in it helps to eliminate bugs in the program by narrowing the possibilities of use.

Here are the most common data types in VB:

- `Integer`: This data type represents positive and negative whole numbers, with a range of -2,147,483,648 to 2,147,483,647
- `Long`: This data type represents more significant whole numbers, with a range of -9,223,372,036,854,775,808 to 9,223,372,036,854,775,807
- `Single`: This data type represents single-precision floating-point numbers, with a range of approximately -3.4×10^{38} to 3.4×10^{38}
- `Double`: This data type represents double-precision floating-point numbers, ranging from approximately -1.7×10^{308} to 1.7×10^{308}
- `Decimal`: This data type represents decimal numbers with fixed precision and a scale ranging from approximately $+/- 7.9 \times 10^{28}$
- `Boolean`: This data type represents a logical value, either `true` or `false`
- `Char`: This data type represents a single Unicode character
- `String`: This data type represents a sequence of Unicode characters
- `Date`: This data type represents a date and time value
- `Object`: This data type represents a reference to an object

In addition to these basic data types, many VB family members support arrays, structures, and enumerations. VB also allows for creating user-defined data types using the `Type` keyword.

Next, we'll discuss the difference between implicit and explicit variable declaration.

Implicit versus explicit variable declaration

In VB, you can use explicit variable declaration to specify the data type of a variable. The precise type declaration contrasts the implicit variable declaration, where VB automatically assigns a data type to a variable based on the initial value.

To use explicit variable declaration, you can use the `As` keyword to specify the data type of a variable when you declare it. Here's an example:

```
Dim myInteger As Integer
myInteger = 45
```

The `myInteger` variable is explicitly declared as an `Integer` data type in this example. The declaration means that VB will not allow any other data types to be assigned to `myInteger`. When the variable is assigned a value of 45, VB knows that it is an integer value and can perform integer arithmetic.

By default, VB uses implicit variable declaration if you do not specify a data type. Here's an example:

```
Dim myString = "Hello, world!"
```

The `myString` variable is not explicitly declared with a data type in this example. However, because it is initialized with a string value, VB will assign the `String` data type to it.

The implicit variable declaration can be convenient, but it can also lead to errors if the wrong data type is assigned to a variable. The explicit variable declaration can help prevent these errors by ensuring that a variable always has the correct data type.

Exploring the Option Explicit statement

`Option Explicit` is a statement in VB that requires all variables to be explicitly declared before they are used in the code. When `Option Explicit` is enabled, VB will not allow implicit variable declaration, which can help prevent typographical errors and other mistakes in the code.

To enable `Option Explicit`, you must include the statement at the beginning of the module before any other code. Here's an example:

```
Option Explicit

Sub mySub()
   myInteger = 45 ' Error: Variable not declared.
   Dim myInteger As String ' Error: Type mismatch.
   myInteger = "Hello, world!"
   Debug.Print myInteger
End Sub
```

In this example, the first line generates an error because `myInteger` has not been declared before it is used. The second line causes an error because `myInteger` is declared a `String` data type but assigned an integer value.

By enabling `Option Explicit`, you can catch these types of mistakes before the code is run, which can save time and prevent bugs.

Next, let's look at where a variable can be referenced in your program.

Understanding variable scope

In VB, variable scope refers to where a variable can be accessed within a program. Variables can be declared at different levels of scope, and their visibility and accessibility can depend on where they are declared.

There are three levels of variable scope in VB:

- **Procedure-level scope**: Variables declared within a procedure or function are only accessible within that procedure or function. Once the procedure or function has been completed, the variables are destroyed, and their values are lost. Here's an example:

```
Sub mySub()
   Dim myInteger As Integer
   myInteger = 45
End Sub
```

In this example, the myInteger variable is declared within the mySub procedure. Therefore, it can only be accessed within that procedure and is destroyed when completed.

- **Module-level scope**: Variables declared outside any procedure or function are accessible throughout the entire module. They can be used in any procedure or function within the module. Here's an example:

```
Dim myString As String

Sub mySub()
   myString = "Hello, world!"
End Sub
```

In this example, the myString variable is declared outside any procedure or function. Therefore, it can be accessed within the mySub procedure and any other procedure or function within the module.

- **Global-level scope**: Variables declared outside of any module are accessible throughout the entire project. They can be used in any module within the project. Here's an example:

```
Public myDouble As Double

Sub mySub()
   myDouble = 3.14159
End Sub
```

In this example, the myDouble variable is declared outside of any module. Therefore, it can be accessed within the mySub procedure and any other module within the project.

It's important to note that variables with the same name can be declared at different levels of scope. In this case, the variable with the narrowest scope (that is, the one declared within the procedure or function) takes precedence over variables with the same name declared at wider scopes.

Best practices – naming variables

Variable naming is an essential aspect of programming as it can affect the code's readability, maintainability, and efficiency. Here are some best practices for variable naming:

- **Use descriptive names**: Choose variable names that indicate what the variable represents. For example, use `totalPrice` instead of `tp` or `customerName` instead of `cn`.

- **Use camelCase**: Use camelCase to separate words in variable names. camelCase is a convention where the first word is in lowercase, and subsequent words start with an uppercase letter. So, for example, use `firstName` instead of `first_name`.

- **Avoid using numbers and special characters**: Do not start a variable name with a number or a special character, except for the underscore (_) character. Also, avoid using special characters in variable names, as they can make the code harder to read and understand.

- **Use singular nouns**: Use singular nouns for variable names unless the variable represents a collection or an array. For example, use `book` instead of `books`.

- **Be consistent**: Use consistent naming conventions throughout your code, and make sure that variables with similar meanings are named consistently. For example, use `firstName` and `lastName` instead of `firstName` and `lname`.

- **Use meaningful prefixes or suffixes**: Use prefixes or suffixes to indicate the data type or purpose of the variable. For example, use `str` for string variables, `int` for integer variables, or `lst` for list variables.

- **Use readable names for boolean variables**: Use names that are easy to understand for boolean variables. For example, use `isFound` instead of `flag` or `bool`.

- **Avoid using reserved keywords**: Do not use them as variable names, as they have a special meaning in the programming language. For example, do not use `Sub`, `Function`, or `Dim` as variable names in VB.

By following these best practices for variable naming, you can improve your code's readability, maintainability, and efficiency.

Storing more than one value

An array is a collection of variables of the same data type. In VB, you can create arrays to store multiple values of the same data type in a single variable. Here's how to declare an array in VB:

```
Dim myArray(4) As Integer
```

In this example, an array called myArray is declared to hold five integer values (0 to 4).

You can also assign values to an array at the time of declaration:

```
Dim myArray(2) As String
myArray = {"red", "green", "blue"}
```

In this example, an array called myArray is declared to hold three string values. The values are assigned to the collection at the time of declaration.

To access the values in an array, you use an index number that specifies the position of the value within the array. The first element of the array has an index of 0. Here's an example:

```
Dim myArray(2) As String
myArray = {"red", "green", "blue"}

Console.WriteLine(myArray(0))
Console.WriteLine(myArray(1))
Console.WriteLine(myArray(2))
```

In this example, the values of the myArray array are accessed using their index numbers.

You can also use loops (we will cover loops in detail in *Chapter 6*) to iterate through the values in an array:

```
Dim myArray(2) As String
myArray = {"red", "green", "blue"}

For i = 0 To 2
    Console.WriteLine(myArray(i))
Next
```

In this example, a For loop is used to iterate through the values in the myArray array and then display them to the console.

Arrays in VB can have up to 2,147,483,647 elements (the maximum value of an Integer data type). VB also supports multidimensional arrays, which are arrays with more than one dimension. Multidimensional arrays can help store data in a table format or for work with matrices.

Next, let's look at the small difference in data types among the different family members of VB.

Differences in data types between family members

When it comes to data types, the different VB family members are mostly the same. There are some small differences that I will break down into comparisons in this section.

Differences in data types between VBScript and VB6

Some of the differences in data types between VBScript and VB6 include the following:

- VBScript has a more limited set of data types compared to VB6. For example, VBScript does not have separate data types for `Integer` and `long` integer values and instead uses a single data type called `Integer` to represent both types of values.

- VBScript does not have a separate data type for decimal values. Instead, decimal values are stored as floating-point numbers with limited precision.

- VBScript does not have a `Variant` data type like VB6. Instead, VBScript uses the `Empty` keyword to represent a variable with an undefined value.

- VBScript does not support user-defined data types like VB6. This means you cannot define your data types in VBScript using the `Type` statement.

- VBScript supports the `Null` value, which represents the absence of a value. This value is not supported in VB6.

- VBScript has more limited support for arrays compared to VB6. For example, VBScript does not support multidimensional arrays or user-defined array types.

Overall, VBScript has a more limited and simplified data type system compared to VB6. This makes it easier to write simple scripts and automate tasks but may require some adjustments for developers who are used to the more complex data types in VB6.

Differences in data types between VBA and VB6

Some of the differences in data types between VBA and VB6 include the following:

- VBA and VB6 have similar built-in data types, such as `Boolean`, `Integer`, `Long`, `Single`, `Double`, `Currency`, `Date`, `String`, and `Variant`.

- VBA and VB6 have slightly different rules for declaring variables. In VB6, variables can be declared with the `Dim` statement without an explicit data type, and the data type will be inferred from the variable's context. In VBA, variables must be explicitly declared with a data type using the `Dim` statement.

- VBA supports user-defined data types using the `Type` statement, allowing you to define your data types with named fields. VB6 also supports user-defined data types using the `Type` statement.

- VBA and VB6 have slightly different rules for passing variables between procedures. In VB6, variables are passed by reference by default but can be passed by value using the `ByVal` keyword. In VBA, variables are passed by reference by default but can be passed by value using the `ByVal` keyword or by reference using the `ByRef` keyword.

- VBA has slightly different rules for working with arrays compared to VB6. For example, VBA arrays are always zero-indexed by default, whereas VB6 arrays can be zero-indexed or one-indexed, depending on the `Option Base` statement.

Overall, VBA and VB6 have similar data type systems, but there are some differences in the syntax and rules for declaring and working with variables. Developers familiar with VB6 should be able to learn VBA relatively quickly but may need to adjust to slightly different syntax and rules.

Differences in data types between VB6 and VB.NET

Some of the differences in data types between VB6 and VB.NET include the following:

- The `Variant` data type, used in VB6 to store values of any data type, has been deprecated in VB.NET. Instead, VB.NET has several other data types that can store values of specific data types, such as `Object`, `String`, `Integer`, `Decimal`, `Double`, and so on.

- VB.NET has several new data types not present in VB6, including `Boolean`, `Char`, `Date`, and `Byte`. These data types store values of specific data types, such as `true/false` values, characters, dates, and byte values, respectively.

- In VB6, variables are declared using the `Dim` statement, and their data types are determined by their context. In VB.NET, variables are declared using the `Dim` statement with an explicit data type or the `Option Infer` statement to allow the compiler to infer the data type based on the variable's initial value.

- VB.NET has stricter rules for variable declaration and scope than VB6. For example, variables in VB.NET must be explicitly declared with a data type, and their scope must be explicitly defined using keywords such as `Private`, `Public`, `Friend`, and so on.

- VB.NET supports arrays and collections with strongly typed elements, meaning that an array or collection of elements must be of a specific data type.

- VB.NET has built-in support for enumerations and user-defined data types, allowing you to define a set of named constants.

VB.NET has a more modern and strongly typed data type system compared to VB6, making it easier to write safe and reliable code. However, the changes in data types between the two languages may require some adjustments for developers familiar with VB6.

Summary

This chapter introduced data types and variables in the six environments where Visual Basic programming is relevant. As we progress in this book, we will build on these tools so you can read and write complete programs in each environment. In the next chapter, we will dive into decision branching, allowing you to control the pathways of code that get executed based on runtime information.

4

Decision Branching

Decision branching is a programming construct that allows a program to execute different sets of instructions based on a condition or set of requirements. There are several ways to implement decision branching in programming languages, but the most common methods include If statements, switch or Case statements, and ternary operators.

In this chapter, we're going to cover the following main topics:

- If statements
- Case statements
- Ternary operators
- VB.NET OrElse operator
- VB.NET AndAlso operator

If statements

In Visual Basic, the If statement executes a block of code if a condition is true. The basic syntax of an If statement in Visual Basic is as follows:

```
If condition Then
    ' code to execute if condition is true
End If
```

For example, the following VB6 code checks whether a number is positive:

```
Dim test As Integer = 12
If test > 0 Then
    MsgBox ("The number is positive")
End If
```

You can also use an `Else` statement to execute a different block of code if the condition is false. The basic syntax of an `If-Else` statement in Visual Basic is as follows:

```
If condition Then
    ' code to execute if condition is true
Else
    ' code to execute if condition is false
End If
```

For example, the following VB6 code checks whether a number is positive or negative:

```
Dim test As Integer = -12
If test > 0 Then
    MsgBox("The number is positive")
Else
    MsgBox("The number is negative")
End If
```

You can also use `ElseIf` statements to check additional conditions. The basic syntax of an `If-Elseif-Else` statement in Visual Basic is as follows:

```
If condition1 Then
    ' code to execute if condition1 is true
ElseIf condition2 Then
    ' code to execute if condition2 is true
Else
    ' code to execute if neither condition1 nor condition2 is true
End If
```

For example, the following VB6 code checks whether a value is positive, negative, or zero:

```
Dim test As Integer = 0
If test > 0 Then
    MsgBox("The number is positive")
ElseIf test < 0 Then
    MsgBox("The number is negative")
Else
    MsgBox("The number is zero")
End If
```

Next up, we will look at an alternative approach to modifying the flow of the program with `Case` statements.

Case statements

A Case statement in VB is used to perform different actions based on the value of a variable or expression. It allows you to evaluate an expression against multiple possible values and execute different code depending on which value the expression matches.

Here's the basic syntax of a Case statement in VB:

```
Select Case expression
    Case value1
        ' Code to execute if expression matches value1
    Case value2
        ' Code to execute if expression matches value2
    Case value3
        ' Code to execute if expression matches value3
    Case Else
        ' Code to execute if expression doesn't match any of the
          above values
End Select
```

Let's see an example of how to use a Case statement in VB.NET:

```
Dim var1 As Integer
var1 = 3
Select Case var1
    Case 1
        Console.WriteLine("The number is one.")
    Case 2
        Console.WriteLine("The number is two.")
    Case 3
        Console.WriteLine("The number is three.")
    Case Else
        Console.WriteLine("The number is not 1, 2, or 3.")
End Select
```

In the preceding example, the var1 variable is assigned a value of 3; a Case statement is used to check the value of var1 and execute the appropriate code block. Since var1 equals 3, the code block for Case 3 is performed, which outputs "The number is three." to the console.

Next, we will look at a particular operator allowing conditional statements within an expression.

Ternary operators

In VB, the ternary operator allows you to write more concise code for conditional statements that evaluate a single value. The ternary operator is a shorthand way of writing an `If-Then-Else` statement that assigns a value to a variable based on a condition.

The syntax of the ternary operator in VB is as follows:

```
variable = If(condition, trueValue, falseValue)
```

In this syntax, `variable` is the name of the variable to which the result of the ternary operator is assigned, `condition` is the condition that is evaluated, `trueValue` is the value set to `variable` if the condition is true, and `falseValue` is the value assigned to `variable` if the condition is false.

Here's an example of how to use the ternary operator in VB.NET:

```
Dim var1 As Integer = 5
Dim var2 As Integer = 10

Dim largestNum As Integer = If(var1 > var2, var1, var2)
Console.WriteLine("The largest number is: " & largestNum)
```

In this example, the ternary operator assigns the larger of two numbers, `var1` and `var2`, to the `largestNum` variable. If `var1` is greater than `var2`, `var1` is assigned to `largestNum`, otherwise, `var2` is assigned to `largerNum`. Finally, the largest number is output to the console using the `Console.WriteLine` method.

We will now move on to a new operator for VB.NET that allows faster execution.

The VB.NET OrElse operator

In VB.NET and ASP.NET, the `OrElse` operator is a logical operator used with `If` statements to evaluate multiple conditions. The `OrElse` operator is like the `Or` operator but provides short-circuit evaluation. Short-circuit evaluation means the testing can stop after one condition evaluates to true.

The basic syntax for using the `OrElse` operator in an `If` statement is as follows:

```
If condition1 OrElse condition2 Then
    ' code to execute if either condition is true
End If
```

In the preceding example, the code block will be executed if either `condition1` or `condition2` is true. However, unlike the `Or` operator, the `OrElse` operator provides short-circuit evaluation, which means that if `condition1` is true, `condition2` will not be evaluated. Short-circuiting can improve performance when considering `condition2` is expensive or time-consuming.

Here's an example that demonstrates the use of the `OrElse` operator:

```
Dim var1 As Integer = 4
Dim var2 As Integer = 8
If var1 > 4 OrElse var2 < 8 Then
    ' This code block will not be run.
End If
If var1 >= 4 OrElse var2 < 4 Then
    ' This code block will be executed.
End if
```

In this example, the first `If` statement will not execute the code block because *condition1* (`var1 > 4`) is false. The second `If` statement will execute the code block because *condition1* (`var1 >= 4`) is true. Because the `OrElse` operator is used in both `If` statements, if *condition1* is true, *condition2* (`var2 < 4`) will not be evaluated, which can improve performance. The performance is improved by only comparing one condition and moving on to the next statement since it knows the result of the complete expression with just one comparison.

We will now move on to a new operator for VB.NET that allows faster execution for multiple conditions that must be true.

The VB.NET AndAlso operator

In VB.NET, the `AndAlso` operator is a logical operator used with `If` statements to evaluate multiple conditions. The `AndAlso` operator is similar to the `And` operator, but it provides short-circuit evaluation.

The basic syntax for using the `AndAlso` operator in an `If` statement is as follows:

```
If condition1 AndAlso condition2 Then
    ' code to execute if both conditions are true
End If
```

In the preceding example, the code block will be executed if both `condition1` and `condition2` are true. However, unlike the `And` operator, the `AndAlso` operator provides short-circuit evaluation, which means that if `condition1` is false, `condition2` will not be evaluated. Short-circuiting can improve performance in situations where considering `condition2` is expensive or time-consuming.

Here's an example that demonstrates the use of the `AndAlso` operator:

```
Dim var1 As Integer = 4
Dim var2 As Integer = 8

If var1 >= 4 AndAlso var2 < 8 Then
    ' This code block will not be run.
End If
```

```
If var1 >= 4 AndAlso var2 <= 8 Then
    ' This code block will be run.
End If
```

In this example, the first `If` statement will not execute the code block because *condition2* (`var2 < 8`) is false. The second `If` statement will execute the code block because both conditions (`var1 >= 4` and `var2 <= 8`) are true. Because the `AndAlso` operator is used in both `If` statements, if *condition1* is false, *condition2* will not be evaluated, which can improve performance.

The GoTo statement

In VB, the `GoTo` statement transfers control to a specific line or label within your code. While the `GoTo` statement is generally discouraged in modern programming practices, it is still supported in Visual Basic for specific scenarios.

The basic syntax of the `GoTo` statement in Visual Basic is as follows:

```
goto line_or_label
```

Here, `line_or_label` refers to the line number or label you want to transfer control to. To see `GoTo` usage, consider the following example:

```
Sub GotoProc()
    Dim num1 As Integer = 20

    If num1 > 10 Then
        GoTo Label1
    Else
        GoTo Label2
    End If

Label1:
    Console.WriteLine("The number is above 10.")
    GoTo EndLabel

Label2:
    Console.WriteLine("The number is equal or under 10.")
    GoTo EndLabel

EndLabel:
    Console.WriteLine("Program execution finished.")
End Sub
```

In this example, we have a subroutine called `GotoProc` that uses the `GoTo` statement to transfer control based on the value of the `num1` variable. If `num1` is greater than `10`, the control is transferred to `Label1` where a message is displayed. Otherwise, control is transferred to `Label2` where a different message is displayed. Control eventually reaches `EndLabel`, where a completion message is displayed in both cases.

The history of the `GoTo` statement is closely related to decision branching. The `GoTo` statement is a control transfer statement that allows the program execution to jump to a different part of the code. It was a fundamental construct in early programming languages such as **FORTRAN** and **BASIC**.

However, the extensive use of `GoTo` statements led to unstructured and hard-to-maintain code. As a result, programmers often used `GoTo` statements to create spaghetti code, where the program flow became unpredictable and challenging to understand. This practice came to be known as "`GoTo` spaghetti."

In the late 1960s and early 1970s, structured programming emerged as a discipline to address the issues caused by the uncontrolled use of `GoTo` statements. Structured programming emphasizes using structured control flow constructs such as loops and conditional statements to enhance code readability and maintainability.

The introduction of structured programming led to a movement against `GoTo` statements. Influential computer scientists such as Edsger Dijkstra argued that `GoTo` statements should be avoided because they could lead to unmanageable code. Dijkstra's famous paper, *"Goto Statement Considered Harmful,"* published in 1968, significantly changed the perception of `GoTo` statements.

As a result, modern programming languages and coding practices discourage or even eliminate the use of `GoTo` statements. Instead, structured control flow constructs such as conditional statements, loops, and subroutines are used to achieve decision branching in a more structured and manageable manner. In addition, these constructs provide greater clarity and maintainability, making it easier for programmers to reason about the behavior of their code.

Overall, decision branching is an essential concept in programming that allows for the execution of different code paths based on conditions. However, the history of the `GoTo` statement demonstrates how programming languages and practices have evolved to promote structured control flow and discourage the uncontrolled use of `GoTo` statements.

Summary

This chapter introduced decision branching with `If` statements, `Case` statements, and ternary operators. Decision branching is used to execute blocks of code based on conditional testing. This allows different pathways in the code based on input or external conditions. In the next chapter, we will continue our journey with iteration.

5
Iteration

In programming, iteration refers to the process of repeating a sequence of instructions multiple times. A programming language that supports iteration provides a way to execute a block of code repeatedly, either a fixed number of times or until a specific condition is met.

Most programming languages provide some form of iteration control structure, such as a `for` loop, `while` loop, `do-while` loop, or `foreach` loop, that allows programmers to perform iterations in a structured way.

VB is a programming language that supports various forms of iteration, including `For` loops, `While` loops, and `Do` loops. These loops allow you to repeat a block of code a specific number of times or until a particular condition is met. VB also supports statements to change the loop flow while inside the iteration.

In this chapter, we're going to cover the following main topics:

- `For` loops
- `While` loops
- `Do` loops
- `Exit` and `Continue` statements
- Type safety differences in loops between VB.NET and other family members
- Infinite loops

For loops

A For loop is a control structure that allows you to execute a block of code repeatedly for a fixed number of times. The basic syntax for a For loop in VB is as follows:

```
For cntr As Integer = start To end Step stepValue
    'code to execute
Next
```

In this syntax, cntr is the loop variable that is used to control the loop. The loop will start with the value of start and end with the value of end, incrementing the variable by stepValue each iteration the loop executes. The stepValue is optional and defaults to 1 if not specified.

Here's an example of VB.NET code that will loop over a range of numbers and prints each number to the console:

```
For x As Integer = 1 To 7
    Console.WriteLine(x)
Next
```

This code will assign the numbers 1 through 7 to the variable x and then print it to the console.

A For loop can also have a custom step value, for example:

```
For cntr As Integer = 1 To 20 Step 2
    Console.WriteLine(cntr)
Next
```

This code will print the odd numbers from 1 to 20 by incrementing the loop cntr variable by two each time the loop executes. The numbers include the numbers 1,3,5, 7, 9, 11, 13, 15, 17, and 19.

Next, we will look at loops executing zero or more times based on a conditional test.

While loops

In computer programming, a loop with a pre-condition is a control flow structure that repeats a set of statements while a specific condition is true before executing the loop body. This loop type is called a While loop because the condition is checked before running the loop body. Since the loop may never execute, we say a pre-condition loop will always execute zero or more times.

While loops with pre-conditions are helpful when you want to repeat a block of code until a specific condition is no longer true. While loops are often used in situations where the number of times you need to repeat the loop is not known in advance. Usually, this is because the code is dependent on receiving specific input.

The basic syntax for a `While` loop in VB is as follows:

```
While test_condition
    'code to run
End While
```

Here, `test_condition` is the `Boolean` expression evaluated before each loop iteration. The loop will continue executing the code block if `test_condition` is true. When `test_condition` becomes false, the loop terminates, and execution continues with the statement following the loop.

The following is an example of a `While` loop in VB. This example iterates over the numbers 1 to 20 and prints each number to the console:

```
Dim x As Integer = 1
While x <= 20
    Console.WriteLine(x)
    x += 1
End While
```

Next, we will look at loops executing one or more times based on a conditional test.

Do loops

A post-condition loop is a control flow structure in computer programming that repeatedly executes a code block until a specific condition is met after completing the loop body. In contrast to a pre-condition loop, a post-condition loop executes the loop body first and then evaluates the condition to determine whether it should continue iterating. Because of the first iteration, a post-condition loop will always run one or more times.

In VB, a `Do` loop is a control structure that implements a post-condition loop. The basic syntax for a `Do` loop in VB is as follows:

```
Do
    'code to run
Loop While test_condition
```

In this example, the loop will run the code block if `test_condition` is true. After executing the loop body, the `Do` loop evaluates `test_condition` at the end of each iteration. If `test_condition` is false, the loop terminates, and execution continues with the statement following the loop.

Here's a VB.NET example of a `Do` loop in VB that reads user input until a valid integer is entered:

```
Dim sentinel As Boolean = False
Console.WriteLine("Enter test scores. Enter -1 to stop.")
Do
    Dim input As String = Console.ReadLine()
```

```
        If input = "-1" Then
            sentinel = True
        End If
Loop While Not sentinel
```

A sentinel in programming is a particular value or character that signals the end of a sequence or input data. A sentinel is used to mark the end of a list when the length of the list is not known at the start of the loop. The sentinel value is typically chosen not to be mistaken for a valid data value.

In the preceding example, the program reads a sequence of integers from the user, terminated by the value -1. The program uses -1 as the sentinel value to indicate the end of the input.

Next, we will look at ways to modify how the loop executes.

Exit and Continue statements

In VB, the Exit and Continue statements are used to modify the flow of execution in loops and conditional statements.

The Exit statement immediately exits a loop or a Select Case or Do Select statement. When the Exit statement is reached, control passes to the next statement after the loop or the Select Case or Do Select statement.

Here is a VB.NET example that searches for a specific value in a loop and will exit the loop if found:

```
Dim test_value As Integer = 7
Dim found As Boolean = False
For x As Integer = 1 To 20
    If x = test_value Then
        Found = True
        Exit For
    End If
Next
If found Then
    Console.WriteLine("The search value was found.")
End If
```

The Continue statement is used to skip to the next loop iteration immediately. When the Continue statement is executed, the loop jumps directly to the next iteration without executing any statements that follow it in the loop body.

Here is a VB.NET example that calculates the sum of all odd numbers in a loop:

```
Dim sum As Integer = 0
For x As Integer = 1 to 20
    If x Mod 2 = 0 Then
```

```
      Continue For
   End If
   sum += x
Next
Console.WriteLine("Sum of even numbers: " & sum)
```

In this example, the loop utilizes the Mod operator to test whether the number is evenly divisible by 2. If it is, the Continue statement will skip to the next iteration.

Next up, we think about some changes in VB.NET with loops.

> **Type safety differences in loops between VB.NET and other family members**
>
> VB.NET is a strongly typed language, meaning you need to declare the data type of variables. The type is not necessary for older VB family members. For example, in VB.NET, you would declare the counter variable in a For loop like this:
>
> `For counter As Integer = start To end Step increment`
>
> In other VB family members, you would use the following:
>
> `For counter = start To end Step increment`

Infinite loops

An infinite loop is a loop that continues indefinitely without a condition that can make it terminate. The endless loop can happen unintentionally or as part of the program's logic. Infinite loops can cause a program to become unresponsive, consume excessive system resources, or crash.

Here's an example of an infinite loop in VB:

```
While 1=1
    'code to run
End While
```

In this example, the 1=1 condition always evaluates to true so that the loop will continue indefinitely. To break out of this loop, you would need to manually interrupt the program execution, such as by pressing *Ctrl + C*.

While infinite loops are usually unintended and considered bugs, there are cases where intentionally creating infinite loops can be helpful. For instance, an infinite loop can be used in server applications to listen for and handle client requests continuously.

It's crucial to ensure that the loop condition is appropriately defined and will eventually evaluate to a false value to avoid unintended infinite loops. Additionally, using loop control statements such as exit or return inside the loop can provide a way to exit the loop under certain conditions.

Summary

This chapter introduced iteration with For, While, and Do loops. Iteration is used to execute blocks of code multiple times. We also discussed special statements to break out of loops. In the next chapter, we will continue our journey with a look at modularizing our code to make it easier to reuse code.

6

Functions and Procedures

Procedures and functions are essential concepts in programming that allow developers to break down complex tasks into smaller, manageable pieces of code that can be reused throughout a program.

A procedure is a named section of code that performs a specific task. The procedure does not return a value and typically modifies data within the program. Procedures are used to group related tasks, such as input/output operations or data validation, and can be called from other program parts.

A function is a named section of code that returns a value. It may or may not modify data within the program. Functions are used to perform specific calculations or operations and can be called from other parts of the program.

Both procedures and functions can take arguments; inputs are passed to the code block to perform the task. For example, arguments can be used to provide a code block with necessary data or modify the behavior of the code block.

In this chapter, we will cover the following main topics:

- Modularizing code in functions
- Modularizing code in procedures
- Understanding parameters in functions and procedures

Modularizing code in functions

VB provides hundreds of built-in functions. These functions can be used to perform standard calculations, manipulate strings and arrays, and interact with the user interface of a program. These out-of-the-box functions are basically code-written and tested for you to reuse. Here are some commonly used functions in VB:

- `Abs`: Returns the absolute value of a number without its sign.
- `Date`: Returns the current date while a program is running.
- `Cint`: Converts a parameter to an integer.

- `Format`: Formats a numeric or date value as a string.

- `InStr`: Returns the position of the first occurrence of a substring in a string. This is similar to the `find` functionality in word processing software.

- `Int`: Returns the integer portion of a number.

- `Lcase`: Converts a string to lowercase.

- `Left`: Extracts a specified number of characters from the beginning of a string. This is similar to the `Mid` function but only works from the left side of the string.

- `Len`: Returns the length of a string. Each character in the string counts as one value.

- `Mid`: Extracts a specified number of characters from a string, starting at a fixed position. This is similar to the `Left` and `Right` functions but can start from any place in the string.

- `MsgBox`: Displays a message box with a message and optional buttons.

- `Now`: Returns the current date and time while a program executes.

- `Replace`: Replaces one substring with another in a string.

- `InStr`: Returns the position of the first occurrence of a substring in a string. This is similar to the `replace` functionality in word processing software.

- `Right`: Extracts a specified number of characters from the end of a string. This is similar to the `Mid` function but only works from the right side of the string.

- `Rnd`: Returns a random number between zero and one. The value can be multiplied by an integer to create larger random values.

- `Time`: Returns the current time.

- `Trim`: Removes leading and trailing spaces from a string.

- `Ucase`: Converts a string to uppercase.

In addition to the built-in functions, VB allows developers to create custom functions to perform specific tasks within their programs. Custom functions are defined using the `Function` keyword, followed by the name of the function, any arguments it takes (if any), and the code that should be executed when the function is called. Here's an example:

```
Function mySum(n1 As Integer, n2 As Integer) As Integer
    Dim s As Integer
    s = n1 + n2
    mySum = s
End Function
```

This function takes two integer arguments (n1 and n2), calculates their sum, and returns the result as an integer. The As Integer statement indicates that the function returns an integer value.

To call the function, you can use its name and pass in the required arguments:

```
Dim s As Integer
s = mySum(7, 8)
MsgBox("The sum of the two numbers is " & s)
```

In this example, the resulting variable is assigned the value returned by the mySum function when passed the arguments, 7 and 8. The MsgBox statement displays a message box with the result of the calculation.

Custom functions can be used to encapsulate complex calculations or operations and called from any part of a program when needed. As a result, developers can write code that is easier to read, understand, and maintain by defining custom functions.

Next, we will look at how to modularize code blocks utilizing **procedures**.

Modularizing code in procedures

In addition to custom functions, VB allows developers to create custom procedures to perform specific tasks within their programs. Procedures are defined using the Sub keyword, followed by the name of the procedure, any arguments it takes (if any), and the code that should be executed when the procedure is called. Here's an example:

```
Sub dspMsg(msg As String)
    MsgBox msg
End Sub
```

This procedure takes a string argument, msg, and displays it in a message box using the MsgBox function. The Sub keyword indicates that the procedure does not return a value.

To call the procedure, you can use its name and pass in the required arguments:

```
dspMsg("Hello and welcome!")
```

In this example, the DisplayMessage procedure is called with the argument, "Hello and welcome!", displayed in a message box.

Custom procedures can be used to encapsulate complex tasks or sequences of operations and called from any part of the program when they are needed. By defining custom procedures, developers can write code that is easier to read, understand, and maintain. Additionally, procedures can be used to group related tasks, making managing and modifying code easier.

Next, we will examine how parameters are passed and used in functions and procedures.

Understanding parameters in functions and procedures

Parameters are used in VB to pass values or objects to a function or procedure when it is called. Parameters are specified in the function or procedure definition and are used to define the inputs required by the function or procedure. Here's an example of a function with two parameters:

```
Function myMult(n1 As Integer, n2 As Integer) As Integer
    Dim m As Integer
    m = n1 * n2
    myMult = m
End Function
```

In this example, the myMult function takes two integer parameters, n1 and n2. The function uses these parameters to multiply them, which is returned as an integer value.

You must pass in the required values or objects when calling a function or procedure that takes parameters. Here's an example of calling the myMult function:

```
Dim res As Integer
res = myMult(7, 8)
Console.WriteLine("Result: " & res)
*
```

In this example, the myMult function is called with the arguments 7 and 8. These values are passed as the n1 and n2 parameters, respectively. The result of the calculation is stored in the resulting variable.

VB supports several types of parameters, including the following:

- ByRef: Passes a reference to a variable, allowing the called function or procedure to modify its value.

- ByVal: Passes a copy of a variable, allowing the called function or procedure to use its value but not modify it.

- Optional: Allows a parameter to be omitted when calling the function or procedure. Optional parameters are supported in all family members except VBScript.

- ParamArray: Allows various arguments to be passed to a function or procedure. ParamArray parameters are supported by all family members except VBScript.

Using parameters in VB allows for more flexible and reusable code, as functions and procedures can be designed to work with different inputs, depending on the program's needs.

Return statement

In VB, the Return statement is used to exit a function or a Sub procedure and return control to the calling code. It is used to provide a result or value back to the caller if necessary. The Return statement can be used in different contexts, depending on whether you work with a function or a procedure.

Here is an example of using the Return statement in the VB function:

```
Function myMult(n1 As Integer, n2 As Integer) As Integer
    Dim m As Integer
    m = n1 * n2
    Return m
End Function
```

In this example, the Return statement returns the result of the two numbers, n1 and n2, multiplied together to the caller. We saw earlier that a return value can also be set by assigning the function name a value. The second method does not stop execution at that point compared to the Return statement, which will immediately return to the calling code.

Here is an example of using the Return statement in the VB procedure:

```
Sub DisplayMessage()
    Console.WriteLine("Welcome")
    Return
End Sub
```

In this example, the Return statement is used to exit the Sub procedure and return control to the caller without providing a specific value.

It's important to note that the Return statement is not required in either a procedure or a function. If a Return statement is omitted, the control automatically returns to the caller when the procedure or function ends.

In addition to the Return statement, VB also provides the Exit Sub and Exit Function statements, which can be used to exit a Sub procedure or a function early without executing the remaining code.

Summary

This chapter introduced functions and procedures to modularize code for reuse. Once we have modularized our code, we can reuse it without having multiple copies of the code. This allows our code to have less bugs and also allows for more reusable code. In the next chapter, we will utilize the skills we have acquired so far in a project.

7

Project Part I

In our first project, we want to use the skills we have learned to build a simple calculator for each VB family member. The calculator should allow the user to enter a simple mathematical operation such as addition, subtraction, multiplication, or division. The calculator should also enable the user to enter the two operands used in the calculation. The calculator will then show the result and allow for another calculation.

In this chapter, we're going to cover the following main topics:

- Developing a calculator in the VB.NET console
- Developing a calculator in VB6
- Developing a calculator in **Visual Basic for Applications** (**VBA**)
- Developing a calculator in VBScript
- Developing a calculator in Classic ASP
- Developing a calculator in ASP.NET

Technical requirements

Please complete the steps to install and code with the **Visual Basic** (**VB**) family member from *Chapter 1*. The code for this chapter is available in the GitHub repository for this book: https://github.com/PacktPublishing/Learn-Visual-Basics-Quick-Start-Guide-/tree/main/ProjectPartI.

Building the project in the VB.NET console

This simple console application written in VB.NET programming language performs basic arithmetic operations on two input numbers.

The program first imports the System namespace. Then, it defines a module called program, which contains a single Sub called Main that takes an array of String arguments.

Within the `Main` method, the program declares five variables: n1, n2, oper of type `String`, a res of type `String` initialized to an empty string, and a `sentinel` of type `String` initialized to `False`.

The program enters a `Do-While` loop until the sentinel is set to `True`. Within the loop, the program prompts the user to enter the first number or X to quit. If the user enters X, the sentinel variable is set to `True`, and the loop ends. Otherwise, the program prompts the user to enter the second number and the arithmetic operator.

The program then uses a `Select Case` statement to perform the appropriate arithmetic operation based on the user's input. For example, if the operator is +, the program adds the two numbers; if it is -, the program subtracts the second number from the first; if it is *, the program multiplies the two numbers; and if it is /, the program divides the first number by the second, taking care to check for division by zero. Finally, the program displays an error message if the operator is not one of these.

The program then displays the result of the arithmetic operation. The loop continues until the user enters X to quit:

```
Imports System

Module Program
    Sub Main(args As String())
        Dim n1, n2, oper As String
        Dim res As String = ""
        Dim sentinel As String = False
        Do
            Console.WriteLine("Enter the first number or X to quit:")
            n1 = Console.ReadLine()
            If n1 = "X" Then
                sentinel = True
            Else
                Console.WriteLine("Enter the second number:")
                n2 = Console.ReadLine()
                Console.WriteLine("Enter the operator (+, -, *, /):")
                oper = Console.ReadLine()

                Select Case oper
                    Case "+"
                        res = CInt(n1) + CInt(n2)
                    Case "-"
                        res = n1 - n2
                    Case "*"
                        res = n1 * n2
                    Case "/"
                        If n2 = 0 Then
                            Console.WriteLine("Cannot divide by zero!")
```

```
                    Else
                        res = n1 / n2
                    End If
                Case Else
                    Console.WriteLine("Invalid operator!")
            End Select
            Console.WriteLine("The result is: " & res)
        End If
    Loop While Not sentinel = True
  End Sub
End Module
```

Many small changes can be made to make this program nicer. Please try to experiment with ideas to see how you can improve the program.

Coding a Visual Basic.NET console calculator

The following steps allow you to code and execute a Visual Basic.NET Windows Forms version of Hello World:

1. Start **Microsoft Visual Studio**.

2. Choose **Create New Project**.

3. Choose **Console (.NET Framework)** and click **Next**.

4. Accept the default project name and solution name.

5. Change the location if you want your code in a different folder.

6. Accept the default framework.

7. Click **Create Project**.

8. Enter the code example from the preceding section.

9. To run your program, click the green triangle on the toolbar above your code.

Our next section will tackle the calculator example in Visual Basic 6.

Building the project in VB6

We will now build Visual Basic 6 code for a simple calculator program. The program has a form that loads when the application starts. The form contains a loop that prompts the user to input two numbers and an operator and then operates on the two numbers. The loop prompts the user until they enter X to quit. Finally, the program displays the result of the operation in a message box. For example, if the user tries to divide by zero, a message box says, "Cannot divide by zero!".

The program uses the `Cint` function to convert the inputted numbers into integers and handles errors using message boxes:

```
Private Sub Form_Load()
  Dim n1, n2, res, operator, sentinel As String
  sentinel = False
  Do
    n1 = InputBox("Enter the first number or X to quit:")
    If n1 = "X" Then
        sentinel = True
    Else
        n2 = InputBox("Enter the second number:")

        operator = InputBox("Enter the operator (+, -, *, /):")

        Select Case operator
            Case "+"
            res = Cint(n1) + Cint(n2)
            Case "-"
            res = n1 - n2
            Case "*"
            res = n1 * n2
            Case "/"
            If n2 = 0 Then
                MsgBox ("Cannot divide by zero!")
            Else
                res = n1 / n2
            End If
            Case Else
            MsgBox ("Invalid operator!")
        End Select
        MsgBox ("The result is: " & res)
    End If
  Loop While Not sentinel = True
End Sub
```

Again, think of small modifications you can make to this program and give it a try.

Coding a Visual Basic 6 calculator

The following steps allow you to code and execute a Visual Basic 6 version of Hello World:

1. Start **Microsoft Visual Basic 6**.

2. Choose **Standard EXE** as the project type.

3.   When the Designer loads, double-click inside the rectangle representing your form.

4.   Enter the code from the example in the preceding section.

5.   To run your program, click the green triangle on the toolbar above your code.

In our next section, we will tackle the calculator example in VBA.

Building the project in VBA

What follows is VBA code for a simple calculator that allows the user to enter two numbers and an operator and then perform the corresponding arithmetic operation on the two numbers.

The code initializes some variables and then enters a Do-While loop allows the user to perform multiple calculations until they choose to quit. Within the loop, the code displays input boxes for the user to enter the first, second, and operator numbers.

After the user enters the values, the code uses a Select Case statement to check which operator was entered and perform the corresponding calculation. If the user enters an invalid operator or attempts to divide by zero, the code displays a message box with an error message. The result is displayed in a message box if the calculation is successful.

The loop continues until the user enters X to quit, at which point the sentinel variable is set to True, and the loop ends:

```
Option Explicit

Sub VBACalc()

Dim n1, n2, res, operator, sentinel As String
sentinel = False
Do
    n1 = InputBox("Enter the first number or X to quit:")
    If n1 = "X" Then
        sentinel = True
    Else
        n2 = InputBox("Enter the second number:")

        operator = InputBox("Enter the operator (+, -, *, /):")

        Select Case operator
            Case "+"
            res = CInt(n1) + CInt(n2)
            Case "-"
            res = n1 - n2
            Case "*"
```

```
              res = n1 * n2
              Case "/"
              If n2 = 0 Then
                   MsgBox ("Cannot divide by zero!")
              Else
                   res = n1 / n2
              End If
              Case Else
              MsgBox ("Invalid operator!")
         End Select
         MsgBox ("The result is: " & res)
     End If
Loop While Not sentinel = True


End Sub
```

Coding a VBA calculator in Excel

The following steps allow you to code and execute a VBA macro displaying Hello World:

1. Start **Microsoft Excel**.
2. Display the **View** menu.
3. Click the **Macros** button and choose **View Macros** from the drop-down menu.
4. Type the name test as the macro name.
5. Click the **Create** button.
6. When the Designer loads, double-click inside the rectangle representing your form.
7. Enter the code example from the preceding section.
8. Click the green triangle on the toolbar above your code to run your macro.

We will now move on to input and output in VBScript, which is both similar to and different from what you have seen. This will be similar in terms of the flow of the program but different in the way the user input and output are embedded into the VBA host. In this example, the VBA host is Microsoft Excel.

In our next section, we will tackle the calculator example in VBScript.

Building the project in VBScript

The following VBScript program prompts the user to enter two numbers and an operator, performs the specified arithmetic operation, and displays the result.

The Option Explicit statement enforces explicit variable declaration, which helps to prevent errors caused by misspelling or mistyping variable names.

The program uses a loop with a `sentinel` variable to allow the user to enter multiple sets of numbers and operators until they choose to quit. The `sentinel` variable is initially set to `False`.

The program prompts the user to enter the first number or X to quit. If the user enters X, the `sentinel` variable is set to `True`, and the program exits the loop. If the user enters a number, the program prompts them to enter the second number and the operator.

The program then uses a `Select Case` statement to determine which arithmetic operation to perform based on the operator entered by the user. Finally, the program displays an error message if the operator is not one of the specified options (+, -, *, or /).

The program displays an error message if the operator is division (/) and the second number is 0. Otherwise, the program performs the arithmetic operation and shows the result in a message box.

The loop continues until the user enters X to quit:

```vbscript
Option Explicit

Dim n1, n2, res, operator,sentinel
sentinel = False
Do
    n1 = InputBox("Enter the first number or X to quit:")
    If n1 = "X" Then
        sentinel = True
    Else
        n2 = InputBox("Enter the second number:")

        operator = InputBox("Enter the operator (+, -, *, /):")

        Select Case operator
            Case "+"
            res = CInt(n1) + CInt(n2)
            Case "-"
            res = n1 - n2
            Case "*"
            res = n1 * n2
            Case "/"
            If n2 = 0 Then
                MsgBox("Cannot divide by zero!")
            Else
                res = n1 / n2
            End If
            Case Else
            MsgBox("Invalid operator!")
        End Select
```

```
        MsgBox("The result is: " & res)
    End If
Loop While Not sentinel
```

Coding a VBScript calculator with Notepad

The following steps will allow you to code and execute the previous VBScript that runs a calculator program:

1. Start **Notepad**.
2. Enter the code from the preceding section.
3. Save the file as VBScriptCalc.vbs in your root folder.
4. Run CMD to go to a command prompt.
5. Type cscript VBScriptCalc.vbs and press the *Enter* key.

In our next section, we will tackle the calculator example in Classic ASP.

Building the project in Classic ASP

A Classic ASP program follows that prompts the user to enter two numbers and an operator, performs the specified arithmetic operation, and displays the result.

This HTML form allows the user to enter two numbers and an operator, perform the specified arithmetic operation, and display the result.

The form uses the HTTP GET method to send the user's input to the server. The form contains three input fields: n1 for the first number, n2 for the second number, and a drop-down menu, operator, to select the arithmetic operation.

When the user clicks the **Submit** button, the values of the input fields are sent to the server as part of the URL query string.

The code within the <% %> tags is server-side VBScript code that retrieves the values of the input fields from the query string using the Request.QueryString method. It then uses a Select Case statement to determine which arithmetic operation to perform based on the operator selected by the user.

If the operator is division (/) and the second number is 0, the program uses the Response.Write method to display an error message on the web page.

If the length of the first number entered is greater than 0, the program uses the Response.Write method to display the result of the arithmetic operation on the web page.

This code creates a simple web-based calculator that allows users to perform arithmetic operations on two numbers:

```
<!DOCTYPE html>
<html>
<body>
<form method="get">
Enter the first number: <input name="n1">
Enter the second number: <input name="n2">
Enter the operator:
<select name="operator">
<option value="+">+</option>
<option value="-">-</option>
<option value="*">*</option>
<option value="/">/</option>
</select>
<input type="submit">
</form>
<%
Dim n1, n2, res, operator
n1 = Request.QueryString("n1")
n2 = Request.QueryString("n2")
operator = Request.QueryString("operator")
Select Case operator
  Case "+"
    res = CInt(n1) + CInt(n2)
  Case "-"
    res = n1 - n2
  Case "*"
    res = n1 * n2
  Case "/"
    If n2 = 0 Then
        Response.Write("Cannot divide by zero!")
    Else
        res = n1 / n2
    End If
End Select
If len(n1) > 0 then
  Response.Write("The result is: " & res)
End if
%>
</body>
</html>
```

Coding a Classic ASP calculator with Notepad

The following steps will allow you to code and execute a VBScript that displays the calculator developed in this project:

1. Start **Notepad**.
2. Enter all the code from the preceding section.
3. Save the file called `ClassicASPCalc.asp` in the root web directory (this is typically `c:\inetpub\wwwroot`).
4. Open a web browser.
5. Enter this URL: `http://127.0.0.1/ClassicASPCalc.asp`.

In our next section, we will tackle the calculator example in ASP.NET.

Building the project in ASP.NET

What follows is an ASP.NET Web Forms page written in VB.NET that provides a simple UI for users to perform basic arithmetic operations on two input numbers.

The first line is a server-side directive that specifies the page language, event handling, and code-behind file for the page. The `Language` attribute is set to `vb`, which indicates that the page is written in VB.NET. The `AutoEventWireup` attribute is `false`, meaning that the event handlers for the controls on the page will be explicitly defined in the code-behind file. The `CodeBehind` attribute specifies the name and path of the code-behind file for the page. Finally, the `Inherits` attribute specifies the fully qualified name of the class in the code-behind file that provides the event-handling logic for the page.

The rest of the code is HTML markup that defines the UI elements on the page. The `asp:` prefix indicates that the elements are ASP.NET server controls that will be processed on the server side.

The `<form>` element has an `id` attribute of `form1` and a `runat` attribute of `server`, which indicates that the form will be processed on the server side.

Inside the form, there are three text boxes and a drop-down list. The text boxes are defined using the `asp:TextBox` server control and have `id` attributes of `n1` and `n2` and `runat` attributes of `server`, which indicate that they will be processed on the server side. The drop-down list is defined using the `asp:DropDownList` server control and has an `id` attribute of `oper` and a `runat` attribute of `server`. It also contains four `asp:ListItem` elements, which represent the four arithmetic operators that the user can choose from.

Finally, there is an `asp:Button` server control with a `text` attribute of `Submit` and a `runat` attribute of `server`, which indicates that it will be processed on the server side when the user clicks on it:

```
<%@ Page Language="vb" AutoEventWireup="false" CodeBehind="
ASPENTCalc.aspx.vb" Inherits=" ASPENTCalc. ASPENTCalc" %>

<!DOCTYPE html>

<html xmlns="http://www.w3.org/1999/xhtml">
<head runat="server">
    <title></title>
</head>
<body>
    <form id="form1" runat="server">
        <asp:TextBox id="n1" runat="server"/>
        <asp:TextBox id="n2" runat="server"/>
        <asp:DropDownList id="oper" runat="server">
            <asp:ListItem Text="+" Value="+" />
            <asp:ListItem Text="-" Value="-" />
            <asp:ListItem Text="*" Value="*" />
            <asp:ListItem Text="/" Value="/" />

        </asp:DropDownList>
        <asp:Button text="Submit" runat="server"/>
    </form>
</body>
</html>
```

Next is the code-behind file for an ASP.NET Web Forms page written in VB.NET, which provides a simple UI for users to perform basic arithmetic operations on two input numbers.

The code defines a class called `WebForm1`, which inherits from the `System.Web.UI.Page` class and is the base class for all ASP.NET Web Forms pages.

The `Page_Load` event handler is defined and executed when the page is loaded. Inside the event handler, the values of the n1, n2, and oper variables are obtained from the HTTP POST request sent when the user clicks the **Submit** button on the page. The `Request.Form` method retrieves the values of the form fields with the corresponding name attributes.

Suppose the n1 variable is not null, meaning the form has been submitted. In that case, a `Select Case` statement is used to perform the appropriate arithmetic operation based on the value of the oper variable. The result of the operation is stored in the `res` variable.

If the `n1` variable has a length greater than 0, which means that the user has entered a value for the first number, the result of the operation is displayed on the page using the `Response.Write` method.

If the `n1` variable is null, which means that the page has just been loaded and the user has not submitted the form yet, nothing is displayed on the page.

Overall, this code defines the event handler for the page load event, retrieves the input values from the form, performs the arithmetic operation based on the selected operator, and displays the result on the page:

```
Public Class ASPENTCalc
    Inherits System.Web.UI.Page

    Protected Sub Page_Load(ByVal sender As Object, ByVal e As System.
EventArgs) Handles Me.Load
        Dim n1, n2, oper As String
        Dim res As String = ""
        n1 = Request.Form("n1")
        n2 = Request.Form("n2")
        oper = Request.Form("oper")
        If Not n1 Is Nothing Then
            Select Case oper
                Case "+"
                    res = CInt(n1) + CInt(n2)
                Case "-"
                    res = n1 - n2
                Case "*"
                    res = n1 * n2
                Case "/"
                    If n2 = 0 Then
                        Response.Write("Cannot divide by zero!")
                    Else
                        res = n1 / n2
                    End If
            End Select
            If Len(n1) > 0 Then
                Response.Write("The result is: " & res)
            End If
        End If
    End Sub

End Class
```

Coding an ASP.NET calculator with Visual Studio

The following steps will allow you to code and execute an ASP.NET web page that displays Hello World:

1. Start **Microsoft Visual Studio 2019**.
2. Choose **Create New Project**.
3. Choose **ASP.NET Web Application** and click **Next**.
4. Accept the default project name and solution name.
5. Change the location if you want your code in a different folder.
6. Accept the default framework.
7. Click **Create Projec**t.
8. Choose **Web Forms**.
9. When the Designer loads, navigate to the solution explorer and right-click on the solution.
10. Choose **Add Item** from the pop-up menu.
11. Choose **Web Form Visual Basic**.
12. Navigate to the source tab and enter the ASPX code from earlier.
13. Right-click on the ASPX source and choose **View Code** from the pop-up menu.
14. Enter the VB.NET code from the previous code example.
15. To run your program, click the green triangle on the toolbar above your code.

You now have a good example of what we have learned so far about the different members of the Visual Basic family. Try to reflect on each of the topics we have discussed so far and how we utilized them in this chapter.

Summary

In this chapter, we utilized the skills we have gained to build an interactive calculator for each of the members of the VB family. Make sure you can run the code in the VB family member you are interested in. In the next chapter, we will drill into formatting and modifying data.

Part 2:
Visual Basic Files
and Data Structures

In the second part of this book, we will focus on building our skills for more advanced programs. You will learn how to format data and persist it into files on the disk drive. We will also look at advanced data structures that are provided to you in the Visual Basic programming language.

This part has the following chapters:

- *Chapter 8, Formatting and Modifying Data*
- *Chapter 9, File Input and Output*
- *Chapter 10, Collections*
- *Chapter 11, Project Part II*

Formatting and Modifying Data

Formatting data is a common task in programming languages to present data in a desired format. A programmer will also have to convert data in the same format or to a different format to allow it to be used in an algorithm.

In this chapter, we're going to cover the following main topics:

- Modifying string data
- Modifying numerical data
- Modifying date and time data
- Formatting numeric values
- Formatting date and time values

Modifying string data

In VB, you can modify string data using various built-in functions and operators. Here are some standard operations you can perform on string data:

Concatenation

There are a few ways to combine strings together in **VB**:

- The & operator: You can concatenate strings using the & operator. Here is an example:

```
Dim firstName As String = "Sally"
Dim lastName As String = "Stone"
Dim fullName As String = firstName & " " & lastName
```

The preceding code will make one string from the three separate strings Sally, a space, and the last name, Stone. The concatenation operator is available for all VB family members.

- The `String.Concat` method: Another way to concatenate strings is using the `String.Concat` method. Here is the same example utilizing the `String.ConCat` method:

```
Dim firstName As String = "Sally"
Dim lastName As String = "Stone"
Dim fullName As String = String.Concat(firstName, " ",lastName)
```

The preceding code has the same effect as the concatenation operator example. It will make one string from the three separate strings `Sally`, a space, and the last name `Stone`. The `String.Concat` method is only available in the VB.NET family member.

Substring extraction

There are a few ways to pull part of a string out of another string in **VB**:

- The `Mid` function: Using the `Mid` function, you can extract a substring from a string, for example:

```
Dim myStr As String = "Hello World!"
Dim substr As String = Mid(myStr, 6, 5)
```

This example extracts the substring starting at index 6 with a length of 5 to pull out the word `World` and store it in another variable. The last parameter can be left off.

- The `Substring` method: Using the `Substring` method was added to VB.NET so you can extract a substring from a string, for example:

```
Dim myStr As String = "Hello World!"
Dim substr As String = myStr.Substring(6, 5)
```

The preceding example extracts the substring starting at index 6 with a length of 5 to pull out the word `World` and store it in another variable. With VB.NET, you can still utilize the `Mid` function to do the same.

String replacement

Using the `Replace` method, you can replace specific occurrences of a substring within a string. Here is an example in VB.NET:

```
Dim myStr As String = "Hello World!"
Dim repStr As String = myStr.Replace("World", "Sally")
```

The preceding code replaces the word `"World"` with the word `"Sally"`. The `repStr` variable will now hold the string `"Hello Sally!"`

In earlier family members of VB, the `Replace` function takes the string to operate as the first parameter. Here is an example in VBA:

```
Dim myStr As String = "Hello World!"
Dim repStr As String = Replace(myStr,"World", "Sally")
```

This code works just like the VB.NET version and replaces the word `"World"` with the word `"Sally"`. The `repStr` variable will now hold the `"Hello Sally!"` string.

Case conversion

Using the `ToUpper` method, you can convert a string to uppercase. For example, in VB.NET, you can do the following:

```
Dim myStr As String = "hello world!"
Dim uppStr As String = MyStr.ToUpper()
```

This code converts the string to uppercase. The `uppStr` variable will hold the `"HELLO WORLD!"` value.

In earlier family members of VB, there is a similar function named `Ucase`. The `UCase` function takes the string to operate as a parameter. Here is an example in VBA:

```
Dim myStr As String = "hello world!"
Dim uppStr As String = Ucase(MyStr)
```

This code works like the VB.NET version and converts the string to uppercase. The `uppStr` variable will hold the `"HELLO WORLD!"` value.

Using the `ToLower` method, you can convert a string to lowercase. For example, in VB.NET, you can do the following:

```
Dim myStr As String = "Hello World!"
Dim lowStr As String = MyStr.ToLower()
```

The preceding code converts the string to uppercase. The `lowStr` variable will hold the `"hello world!"` value.

In earlier family members of VB, there is a similar function named `LCase`. The `LCase` function takes the string to operate as a parameter. Here is an example in VBA:

```
Dim myStr As String = "hello world!"
Dim lowStr As String = LCase(MyStr)
```

This code works like the VB.NET version and converts the string to lowercase. The `lowStr` variable will hold the `"hello world!"` value.

Next, we will look at how to modify numerical data in VB.

Modifying numerical data

In VB, you can modify numerical data using many different operators and functions. We will focus on three types of numerical operations: mathematical, rounding/truncation, and type conversions.

Mathematical operations

The traditional math operations are supported in **VB**:

- **Addition**: You can add numeric values using the + operator. Here is an example:

```
Dim n1 As Integer = 4
Dim n2 As Integer = 2
Dim n3 As Integer = n1 + n2
```

The preceding code will declare two integer variables (n1 and n2) and add them together, storing their result in the third variable named n3.

- **Subtraction**: You can subtract values using the - operator. Here is an example:

```
Dim n1 As Integer = 4
Dim n2 As Integer = 2
Dim n3 As Integer = n1 - n2
```

The preceding code will declare two integer variables (n1 and n2) and subtract n2 from n1, storing the result in the third variable named n3.

- **Multiplication**: You can multiply values using the * operator. Here is an example:

```
Dim n1 As Integer = 4
Dim n2 As Integer = 2
Dim n3 As Integer = n1 * n2
```

This code will declare two integer variables (n1 and n2) and then multiply them together, storing their result in the third variable named n3.

- **Division**: You can divide values using the / operator. The result is a real number. Here is an example:

```
Dim n1 As Integer = 5
Dim n2 As Integer = 2
Dim n3 As Double = n1 + n2
```

The preceding code will declare two integer variables (n1 and n2) and divide them, storing their result in the third variable named n3 of type Double.

- **Modular division**: We often want the remainder of the division's operation in computer science. This operation is called modular division. You can perform modular divisions using the Mod operator. Here is an example:

```
Dim n1 As Integer = 5
Dim n2 As Integer = 2
Dim n3 As Integer = n1 Mod n2
```

The preceding code will declare two integer variables (n1 and n2) and then divide them, storing the remainder of one in the third variable named n3.

Rounding and truncation

There are several functions that allow you to get a decimal value with less precision:

- **Rounding**: Using the Round function, you can round a numeric value to the nearest integer. In VB.NET, you want to call the Round method in the Math namespace, for example:

```
Dim pi As Double = 3.14159
Dim roundedPi As Double = Math.Round(pi)
```

The preceding code will store the value 3 in the roundedPi variable.

In earlier VB family members, the same can be accomplished without using the Math namespace:

```
Dim pi As Double = 3.14159
Dim roundedPi As Double = Round(pi)
```

In both VB.NET and earlier VB family members, you can also pass a second parameter – the number of fractional digits to round to, for example:

```
Dim pi As Double = 3.14159
Dim roundedPi As Double = Math.Round(pi,4)
```

This code will round to four decimal places, storing the value 3.1416 in the roundedPi variable.

- **Truncation**: You can truncate a numeric value using the truncate function. Truncation removes the remaining digits from the real number. In VB.NET, you want to call the truncate method in the Math namespace, for example:

```
Dim pi As Double = 3.14159
Dim trunPi As Double = Math.Truncate(pi)
```

The preceding code will store the value 3 in the trunPi variable.

In **VBA**, the same can be accomplished without using the `Fix` function without the `Math` namespace:

```
Dim pi As Double
Pi = 3.14159
Dim trunPi As Double
trunPi = Fix(pi)
```

VBScript and VB6 did not have a `truncate` function. There are other functions, such as `Fix` and `Int`, if you want to remove the decimal portion. A trick to using these functions while preserving digits to the right of the decimal is multiplying the number by a multiple of `10` and then dividing the result by the same number. Here is an example in **VBScript**:

```
Dim pi
pi = 3.14159
Dim trunPi
trunPi = Int(pi * 10000) / 10000
```

The preceding code will truncate to four decimal places, storing the value 3.1415 in the `trunPi` variable.

Type conversion

Using casting operators, you can convert numeric values from one data type to another. Here is an example:

```
Dim n1 As Integer = 25
Dim n2 As Double = CDbl(n1)
```

The preceding example converts the value 25 from an integer to a double. There are many different numeric conversion functions in VB. They all start with a C followed by the data type they return, as we saw with `CDbl`, which returned a double. Another example is `Cint`, which returns an `Integer`.

Next, we will examine how to work with date and time data.

Modifying date and time data

In VB, you can modify date and time data using various functions and methods provided by the language. We will focus on creating date and time values, date and time math, and date and time component extraction.

Creating date and time values

You can create a `Date` value using the `DateValue` function or by specifying a date literal in the format *#MM/dd/yyyy#*, for example:

```
Dim dateVal As Date = DateValue("10/05/1967")
```

The preceding code creates a `Date` value representing October 5, 1967. The same thing can be accomplished using the string literal:

```
Dim dateLit As Date = #10/05/1967#
```

You can create a `Time` value using the `TimeValue` function or by specifying a time literal in the format *#hh:mm:ss AM/PM#*, for example:

```
Dim timeVal As Date = TimeValue("11:30:15 AM")
```

This example code creates a `Time` value representing 11:30:15 AM. The same thing can be accomplished using the time literal:

```
Dim timeLit As Date = #11:30:15 AM#
```

Date and time math

You can add or subtract a specific time to a `Date` or `Time` value using the `DateAdd` function, for example:

```
Dim curDate As Date = Date.Now
Dim futDate As Date = DateAdd("d", 7, curDate)
```

This code gets the current date and adds seven days to it, storing the result in the `futDate` variable.

The `DateAdd` function's first parameter is a VB constant to represent the type of value being added to the third parameter. Here are the possible types and the constants used:

- yyyy: Year
- q: Quarter
- m: Month
- y: Day of year
- d: Day
- w: Weekday
- ww: Week
- h: Hour
- n: Minute
- s: Second

Date and time component extraction

Using the respective properties or functions, you can extract specific components (year, month, day, hour, minute, or second) from a `Date` or `Time` value. In VB.NET, a property is available on a `Datetime` value. In VB6, VBScript, and VBA, there is a function with the same name. Here's an example in VB.NET:

```
Dim curDate As Date = Date.Now
Dim curYear As Integer = curDate.Year
Dim curMonth As Integer = curDate.Month
Dim curDay As Integer = curDate.Day
Dim curHour As Integer = curDate.Hour
Dim curMinute As Integer = curDate.Minute
Dim curSecond As Integer = curDate.Second
```

The same example in earlier VB members would look like this:

```
Dim curDate As Date = Now()
Dim curYear As Integer = Year(curDate)
Dim curMonth As Integer = Month(curDate)
Dim curDay As Integer = Day(curDate)
Dim curHour As Integer = Hour(curDate)
Dim curMinute As Integer = Minute(curDate)
Dim curSecond As Integer = Second(curDate)
```

Next, we will look at formatting numeric values.

Formatting numeric values

In VB, you can format data using various functions and methods to present it in the desired format. Here are some commonly used techniques for formatting data in Visual Basic:

In VB.NET, you can use the `ToString` method with format specifiers to format numeric values, for example:

```
Dim pi as double  = 3.14159
Dim strOut As String = pi.ToString("0.00")
```

The preceding code in VB.NET will store the value `4314` in the `strOut` variable.

You can also use string interpolation in VB.NET to format numeric values within a string, for example:

```
Dim pi = 3.14159
Dim strOut As String = $"{pi:0.00}"
```

In early VB family members (VBScript, VBA, and VB6), you can use the `FormatNumber` function to format numbers according to your desired format. Here's an example of how you can do that in VBScript:

```
Dim pi
pi = 3.14159
Dim formattedPi
formattedPi = FormatNumber(pi, 2)
WScript.Echo formattedPi
```

The preceding code will output 3.14 for the value of the `formatedPi` variable. In this example, the `FormatNumber` function takes two parameters: the number you want to format and the number of decimal places you want to display. Finally, it returns a string representing the formatted number.

The `FormatNumber` function also has some optional parameters that you can use to customize the formatting further. For example, you can specify whether to include a thousand separator or a decimal separator and the format for negative numbers.

Next, we will look at formatting date and time values.

Formatting date and time values

You can use the `ToString` method in VB.NET with custom format specifiers to format date and time values, for example:

```
Dim curDate As Date = Date.Now
Dim fDate As String = curDate.ToString("MM/dd/yyyy")
Dim fTime As String = curDate.ToString("hh:mm:ss tt")
```

If the preceding code is run on June 5, 2023, at 11:30 A.M., it will format the date as `"06/05/2023"` and format the time as `"11:30:00 AM"`.

In early VB family members (VBScript, VBA, and VB6), you can use the `FormatDateTime` function to format date and time values according to your desired format. Here's an example of how you can do that in VBScript:

```
Dim curDate
curDate = Date()
Dim fDate
fDate = FormatDateTime(curDate, vbShortDate)
WScript.Echo fDate
```

The `FormatDateTime` function formats the date and time according to the custom format provided.

You can use various predefined constants and custom format strings with the `FormatDateTime` function to achieve the desired formatting for dates and times. The predefined constants are `vbGeneralDate`, `vbLongDate`, `vbShortDate`, `vbLongTime`, and `vbShortTime`. Custom format strings can include multiple placeholders such as `yyyy` for the year, `MM` for the month, `dd` for the day, `hh` for the hour, `mm` for the minute, `ss` for the second, and `AM/PM` for the time indicator.

Summary

This chapter introduced functions and procedures to modularize code for reuse. When formatting data, the syntax varies a little more than other topics based on the VB family member. In the next chapter, we will look at how we can persist data to the filesystem and read that data back into our programs.

9

File Input and Output

Files stored on disk drives allow our programs to persist data across executions, enabling us to develop potent tools to help us get our daily work done. In addition, we can treat files as both text and binary data. We will explore the advantages of using both methods. We will also think about organizing the files we use in our programs in directories.

In this chapter, we're going to cover the following main topics:

- Working with directories and files
- Writing text to files
- Writing binary data to files
- Reading text from files
- Reading binary data from files

Working with directories and files

We will utilize two different methods for **VB.NET** and older VB family members. We will see this pattern throughout this chapter as Microsoft added new file features to the .NET language that you can leverage in VB.NET and **ASP.NET**.

Working with files in VB.NET

In VB.NET, you use the `File` class from the `System.IO` namespace to work with files. The `File` class provides static methods for performing various file-related operations.

To create a file, you can use the `Create` method, which returns a `FileStream` object you can later use to write data to the file:

```
Imports System.IO
File.Create("C:\Path\File.txt")
```

To delete a file, you can use the `Delete` method:

```
Imports System.IO
File.Delete("C:\Path\ File.txt")
```

To check whether a file exists, you use the `Exists` property, which returns a Boolean value indicating whether the file exists or not:

```
Imports System.IO
If File.Exists("C:\Path\File.txt") Then
    Console.Writeline("The file exists")
End If
```

You use the `Copy` method to copy a file from one location to another:

```
Imports System.IO
File.Copy("C:\Path\File.txt", "C:\Path\NewFile.txt")
```

To move or rename a file, you use the `Move` method:

```
Imports System.IO
File.Move("C:\Path\File.txt", "C:\Path\ NewFile.txt")
```

To retrieve the file attributes, use the `GetAttributes` method and the `FileAttributes` class:

```
Imports System.IO
Dim attr As FileAttributes = File.GetAttributes("C:\Path\ File.txt")
If attr.HasFlag(FileAttributes.ReadOnly) Then
    Console.Writeline("The file is read-only")
End If
If attr.HasFlag(FileAttributes.Hidden) Then
    Console.Writeline("The file is hidden")
End If
```

Next, we will look at working with directories in VB.NET.

Working with directories in VB.NET

In VB.NET, you can use the `Directory` class from the `System.IO` namespace to work with directories. The `Directory` class provides static methods for performing various directory-related operations.

To create a directory, you use the `CreateDirectory` method:

```
Imports System.IO
Directory.CreateDirectory("C:\Path\NewDirectory")
```

To delete a directory, you use the `Delete` method. With the second parameter, you can specify whether to delete the directory recursively or not:

```
Imports System.IO
Directory.Delete("C:\Path\Directory", recursive:=True)
```

To check whether a directory exists, you use the `Exists` property, which returns a Boolean value indicating whether the directory exists or not:

```
Imports System.IO
If Directory.Exists("C:\Path\Directory") Then
    Console.Writeline("The directory exists")
End If
```

To move or rename a directory, you can use the `Move` method:

```
Imports System.IO
Directory.Move("C:\Path\OldDir", "C:\Path\NewDir")
```

To retrieve the files or subdirectories within a directory, you use the `GetFiles` and `Get Directories` methods:

```
Imports System.IO
Dim files As String() = Directory.GetFiles("C:\Path\Dir")
Dim dirs As String() = Directory.GetDirectories("C:\Path\Dir")
```

The `GetFiles` and `GetDirectories` methods return collections of values, which we will explore in *Chapter 10*.

To retrieve information about a directory, you can use the `GetDirectoryInfo` method to get a `DirectoryInfo` object. You can use this `DirectoryInfo` object to get files and directories, as in our last example, but from the object:

```
Imports System.IO
Dim dirInfo As DirectoryInfo = My.Computer.
FileSystemGetDirectoryInfo("C:\Path\Dir")
Dim files() As FileInfo = dirInfo.GetFiles()
```

You retrieve various attributes of a directory using the `GetAttributes` method, just as you did for a file previously:

```
Imports System.IO
Dim attr As FileAttributes = File.GetAttributes("C:\Path\Dir")
If attr.HasFlag(FileAttributes.ReadOnly) Then
    Console.Writeline("The directory is read-only")
End If
```

```
If attr.HasFlag(FileAttributes.Hidden) Then
    Console.Writeline("The directory is hidden")
End if
```

Next, we will look at working with files in **VBScript**, **VBA**, and **VB6**.

Working with files in older VB family members

Working with files and directories in older VB family members (VBScript, VBA, and VB6) involves using the **file system object** (**FSO**) provided by the **Windows Script Host**. The FSO allows you to perform many operations on files and directories.

To create a directory with FSO, you use the `CreateFolder` method:

```
Set myFSO = CreateObject("Scripting.FileSystemObject")
myFSO.CreateFolder("C:\Path\Directory")
```

To delete a directory with FSO, you use the `DeleteFolder` method. The second parameter specifies whether to delete the directory recursively or not:

```
Set myFSO = CreateObject("Scripting.FileSystemObject")
myFSO.DeleteFolder("C:\Path\Directory", True)
```

To check whether a directory exists with FSO, you can use the `FolderExists` method, which returns a Boolean value indicating whether the directory exists or not:

```
Set myFSO = CreateObject("Scripting.FileSystemObject")
If myFSO.FolderExists("C:\Path\Directory") Then
    Wscript.Echo "The directory exists"
End If
```

To rename or move a directory with FSO, you use the `MoveFolder` method:

```
Set myFSO = CreateObject("Scripting.FileSystemObject")
myFSO.MoveFolder "C:\Path\OldDir", C:\Path\NewDir"
```

Next, we will look at working with directories in VBScript, VBA, and VB6.

Working with directories in older VB family members

To create a file with FSO, you use the `CreateTextFile` method, which has optional second and third parameters to specify whether the file should be overwritten and whether it should be created in Unicode format:

```
Set myFSO = CreateObject("Scripting.FileSystemObject")
Set myFile = myFSO.CreateTextFile("C:\Path\File.txt", True)
```

To delete a file with FSO, you use the `DeleteFile` method, which has an optional second parameter to specify whether only read-only files should be deleted:

```
Set myFSO = CreateObject("Scripting.FileSystemObject")
myFSO.DeleteFile("C:\Path\File.txt")
```

To check whether a file exists with FSO, you use the `FileExists` method, which returns a Boolean value indicating whether the file exists or not:

```
Set myFSO = CreateObject("Scripting.FileSystemObject")
If myFSO.FileExists("C:\Path\File.txt") Then
    Wscript.Echo "The file exists"
End If
```

To rename or move a file with FSO, you use the `MoveFile` method:

```
Set myFSO = CreateObject("Scripting.FileSystemObject")
myFSO.MoveFile "C:\Path\Old.txt", "C:\Path\New.txt"
```

Next, we will look at writing data in text format into files.

Writing test into files

We will continue to utilize the two different methods for VB.NET and older VB family members. For VB.NET, we will use the `File` class from the `System.IO` namespace to write text to files. For VBScript, VBA, and VB6, we will continue to use the FSO we used in the files and directories section.

The `File` class also provides two methods that make writing text to a file in a single line easy. The `WriteAllText` and `AppendAllText` methods allow you to send the parameter to overwrite or append the file:

```
Imports System.IO
Dim myFile As String = "C:\Path\File.txt"
File.WriteAllText(myFile, "This text overwrites the file")
File.AppendAllText(myFile, "This is appended text.")
```

In VB.NET, the `StreamWriter` class can also write text to a file. First, you create a `StreamWriter` instance, and then use its `Write` or `WriteLine` methods to write text:

```
Imports System.IO
Dim myFile As String = "C:\Path\File.txt"
Using writer As New StreamWriter(myFile)
    writer.Write("Here is some text.")
    writer.WriteLine("This text will end the line.")
End Using
```

In earlier family members of VB, we can utilize the FSO as we did earlier. FSO provides the `CreateTextFile` method that we saw earlier. This method creates a new file or opens an existing file for writing. The `CreateTextFile` method returns a `TextFile` object. The `Write` method of the `TextFile` object is then used to write the text to the file. Finally, we need to make sure we close the file. The `Close` method of the `TextFile` object is used to close the file and release the resources:

```
myFile = "C:\Path\File.txt"
myText = "This text will be written to the file"
Set myFSO = CreateObject("Scripting.FileSystemObject")
Set myTextFile = myFSO.CreateTextFile(myFile, True)
myTextFile.Write myText
MyTextFile.Close
```

The implementation of the previous code snippet will depend on the VB family member. For VBScript, this could be run by placing the code in a script file. For VB and VBA, you would place the code in a larger project.

Writing text in VB6 and VBA

VB6 and VBA have unique methods that are unavailable in VBScript for output operations. For example, you open a file for a specific mode of operation and give it a file number, which needs to be between 1 and 511. Once the file is opened, you can write to it and close it:

```
Open "C:\path\file.txt" For Output As #1
Write #1, "This is the text."
Close #1
```

Next, we will look at reading data in text format from files.

Reading text from files

We will continue to utilize the two different methods for VB.NET and older VB family members. For VB.NET, we will use the `File` class from the `System.IO` namespace to read text from files. For VBScript, VBA, and VB6, we will continue to use the FSO we used in the files and directories section.

The `File` class also provides two methods that make writing text to a file in a single line easy. The `ReadAllText` and `ReadlAllLines` methods pull all the text from the file into a single variable. The difference between the two methods is that the second method pulls the data line by line into an array, while the first does not separate the input file lines:

```
Imports System.IO
Dim myFile As String = "C:\Path\File.txt"
Dim myText As String = File.ReadAllText(myFile)
```

In VB.NET, the `StreamReader` class allows you to read text from a file. First, you create a `StreamReader` instance, and then use its `Read` or `ReadLine` methods to read the text:

```
Imports System.IO
Dim myFile As String = "C:\Path\File.txt"
Dim myLine As String
Using reader As New StreamReader(myFile)
    myLine = reader.ReadLine
End Using
```

Reading text in VB6 and VBA

VB6 and VBA have unique methods that are unavailable in VBScript for input operations. For example, you open a file for a specific mode of operation and give it a file number, which needs to be between 1 and 511. Once the file is opened, you can read from it with the `Input` or `LineInput` statements and close it. `Input` will read to the next delimiter, and `LineInput` will read the text up to the following line:

```
Dim myValue
Open "C:\path\file.txt" For Input As #1
Input #1, myValue
MsgBox myValue
Close #1
```

Next, we will look at writing binary data to files.

Writing binary data to files

We will utilize the two methods for VB.NET and older VB family members. For VB.NET, we will use the `File` class from the `System.IO` namespace to write binary data to files. For VBA and VB6, we will use the file number operations we saw in the last two sections. VBScript can use `ADODB.stream` to write binary data to files, but we will leave it out of this section due to its complexity.

The `File` class also provides a method that makes writing binary data to a file in a single line easy. The `WriteAllBytes` method allows you to send an array of bytes to the file:

```
Imports System.IO
Dim myFile As String = "C:\Path\File.dat"
Dim bData() As Byte = {230, 231, 232, 233, 234, 235}
File.WriteAllBytes(myFile, bData)
```

To write binary data to files in VB.NET, you can also use the `FileStream` class from the `System.IO` namespace. Here's an example of how to write binary data to a file utilizing a `FileStream` object:

```
Imports System.IO
Dim myFile As String = "C:\Path\File.dat"
Dim bData() As Byte = {230, 231, 232, 233, 234, 235}
Using fs As New FileStream(myFile, FileMode.Create, FileAccess.Write)
    fs.Write(bData, 0, bData.Length)
End Using
```

In the preceding example, the file path where we want to write the binary data is specified in the `myFile` variable. The `bData` array represents the binary data you want to write to the file.

The `FileStream` object is created with the `FileMode.Create` file path to create a new file (overwriting it if it exists), and `FileAccess.Write` to write access to the file. Inside the `Using` block, the `Write` method of the `FileStream` object is called to write the binary data to the file. The parameters passed to the `Write` method are the `bData` array, the offset (0 in this case), and the length of the binary data array.

The `Using` block ensures that the `FileStream` object is properly closed and disposed of after completing the writing operation.

Writing binary data in VB6 and VBA

VB6 and VBA have unique methods that are unavailable in VBScript for binary output operations. You can open a file for a specific mode of operation and give it a file number, which needs to be between 1 and 511. Once the file is opened, you can write binary bytes with the `Put` statement:

```
Open "C:\path\file.dat" For Binary Access Write As #1
Put #1,,230
Close #1
```

In the preceding example, the second parameter represents the byte offset to write to. In the example, this is blank, meaning the byte with a value of 230 is written at the current location. VB will maintain a location pointer in the file as the data is written. You can utilize a `seek` operation to move the pointer to a specific location before doing the `put` operation, or you can utilize the second parameter to skip forward.

Next, we will look at reading binary data from files.

Reading binary data from files

We will continue to utilize the two different methods for VB.NET and older VB family members. For VB.NET, we will use the `File` class from the `System.IO` namespace to read binary data from files. For VBA and VB6, we will use the file number operations we saw in the previous sections. VBScript can use `ADODB.stream` to read binary data from files, but we will leave it out of this section due to its complexity.

The `File` class also provides a method that makes reading binary data from a file in a single line easy. The `ReadAllBytes` method allows you to receive an array of bytes representing the data from the file:

```
Imports System.IO
Dim myFile As String = "C:\Path\File.dat"
Dim bData() As Byte
bData = File.ReadAllBytes(myFile)
```

To read binary data from files in VB.NET, you can also use the `FileStream` class from the `System.IO` namespace. Here's an example of how to read binary data from a file with a `FileStream` object:

```
Imports System.IO
Dim myFile As String = "C:\Path\File.dat"
Dim bData()
Using fs As New FileStream(myFile, FileMode.Open, FileAccess.Read)
        ReDim bData(fs.Length - 1) As Byte
            fs.Read(bData, 0, bData.Length)
End Using
```

This example specifies the file path from which we want to read the binary data with the `myFile` variable. The `FileStream` object is created with the `FileMode.Open` file path to open an existing file and `FileAccess.Read` to specify read access to the file. Inside the `Using` block, the `bData` byte array is created with a length equal to the size of the file. The `Read` method of the `FileStream` object is then called to read the binary data from the file into the `bData` array.

After reading the binary data, we can process it within the `Using` block. The `Using` block ensures that the `FileStream` object is properly closed and disposed of after completing the reading operation.

Reading binary data in VB6 and VBA

VB6 and VBA have unique methods that are unavailable in VBScript for binary input operations. You can open a file for a specific mode of operation and give it a file number, which needs to be between 1 and 511. Once the file is opened, you can read binary bytes with the `Get` statement:

```
Dim myByte as Byte
Open "C:\path\file.dat" For Binary Access Read As #1
Get #1,,myByte
Close #1
```

In the preceding example, the second parameter represents the `Byte` offset to read from. In the example, this is blank, which means `Byte` is read from the current location.

Summary

This chapter introduced reading and writing data from files. We looked at working with both text and binary data. Storing data in files and reading that data from files is important, as we want our programs to have the ability to persist the work they have done between executions of the program.

In the next chapter, we will look at how to store many values in a single variable using the many collections available in the Visual Basic language.

10
Collections

In programming, collections refer to data structures or containers that are used to store and organize groups of related values or objects. These collections provide various methods and operations that you can use efficiently to manipulate and access the data they contain.

Several common types of collections are used in computer programming, including the following:

- **Arrays**: An array is a fundamental collection type that stores a fixed-size sequence of elements of the same data type. Elements in an array are accessed by their index, which represents their position in the array.

- **Lists**: These are dynamic collections that can grow or shrink in size. They allow for efficient insertion, deletion, and retrieval of elements. Lists can store pieces of different types, often providing additional operations such as sorting and searching.

- **Sets**: These are collections that store unique elements without any particular order. They ensure that each piece appears only once, eliminating duplicates. Sets often provide operations for union, intersection, and difference between sets.

- **Dictionaries**: Also known as maps or associative arrays, dictionaries store key-value pairs. They allow you to efficiently look up values based on their associated keys. Dictionaries provide operations for inserting, deleting, and retrieving values based on their keys.

- **Queues**: These follow a **First-In, First-Out** (**FIFO**) order, where elements are added at the end and removed from the front. Queues are commonly used when the processing order is essential, such as scheduling tasks or processing messages.

- **Stacks**: These follow a **Last-In, First-Out** (**LIFO**) order, where elements are added and removed from the same end. Stacks are often used in scenarios such as parsing expressions, managing function calls, or implementing undo/redo functionality.

- **Linked lists**: These contain nodes that hold data and references to the next node in the sequence. They provide efficient insertion and deletion at any position in the list, but accessing elements via an index is less efficient than using arrays.

- **Tuples**: These are ordered collections that can store elements of different types. Unlike arrays, tuples are typically immutable, meaning their elements cannot be modified after creation.

- **Bags or multisets**: These are like sets but allow duplicate elements. They store a collection of pieces without enforcing uniqueness.

There are solutions for each collection type in VB, but we will focus on the core collections needed and used in a typical program.

In this chapter, we're going to cover the following main topics:

- Working with one-dimensional arrays

- Utilizing multidimensional arrays

- When and how to use lists

- Storing data in a dictionary

Working with one-dimensional arrays

In **VB**, one-dimensional arrays are declared and stored in a linear sequence to store a collection of elements of the same data type. Here's the syntax for declaring a one-dimensional array in VB:

```
Dim arrayName(size) As DataType
```

An example of an array of 10 integers would look like this:

```
Dim myInts(10) As Integer
```

The preceding code will create a variable named myInts, which is an array with 10 elements. In **VB**, like most programming languages, the elements are indexed from 0 to size-1. In this example, they would be indexed from 0 to 9.

Array elements in **VB** are accessed using their index, starting from 0. You can retrieve or assign values to array elements as follows:

```
Dim value As Integer = myInts(3)
myInts(3) = 5
```

This code example accesses the fourth element. First, it retrieves the value and then assigns a new value to the fourth element.

You can determine the length (number of elements) of a one-dimensional array using the Length property:

```
Dim len As Integer = myInts.Length
```

You can iterate over the elements of an array using various looping constructs, such as a `For` loop, as we learned previously in this book:

```
For ind As Integer = 0 To myInts.Length - 1
    Console.WriteLine(myInts(ind))
Next
```

The preceding code will output each element in the `myInts` array using a `For` loop variable named `ind`. The `ind` variable will be assigned a value of `0` up to the length `-1`. The `ind` variable is then used as an index in the `myInts` variable.

There is also a specific loop construct in VB to make this easier called, the `For Each` loop:

```
For Each num in myInts
    Console.WriteLine(num)
Next
```

This code will output each element in the `myInts` array, but it does so without indexing into the array. The `For Each` loop will copy each value in the sequence into the variable after the `For Each` keywords. In this case, the `num` variable will get each value from the `myInts` array.

Next, we will look at working with multidimensional arrays in **VB**.

Utilizing multidimensional arrays

Multidimensional arrays allow you to store and organize data in a spreadsheet-like structure with multiple dimensions. You can have arrays with two or more dimensions, such as two-dimensional arrays, three-dimensional arrays, and so on. Here's an example of how you can declare an array with multiple dimensions:

```
Dim arrayName(size1, size2, ...) As DataType
```

For example, to declare and initialize a two-dimensional array of integers with five rows and 10 columns, you can use the following code:

```
Dim myMatrix(5, 10) As Integer
```

Array elements in multidimensional arrays are accessed using their indices for each dimension. You can retrieve or assign values to array elements as follows:

```
Dim value As Integer = myMatrix(4, 9)
myMatrix(4, 9) = 25
```

The previous code example accesses the element at the fourth row and tenth column. First, it retrieves the value, then assigns a new value to that cell of the multidimensional array.

To get the length or size of each dimension in a multidimensional array, you can use the `GetLength` method while passing in the dimension you want to know the size of:

```
Dim rows As Integer = myMatrix.GetLength(0)
Dim columns As Integer = myMatrix.GetLength(1)
```

You can iterate over the elements of an array using various looping constructs inside other looping constructs, such as a `For` loop, as we learned in *Chapter 5*:

```
For row As Integer = 0 To myMatrix.GetLength(0) - 1
    For col As Integer = 0 To myMatrix.GetLength(1) - 1
        Console.WriteLine(myMatrix(row,col))
    Next
Next
```

Using two loops, the previous code will output each element in the multidimensional `myMatrix` array. The outer loop goes through the first dimension, assigning the index to a variable for the row. The inner `For` loop goes through the second dimension, assigning the index to a variable for the column.

Handling arrays in early VB family members

There is one difference with arrays in early **VB** family members from the VB.NET examples that we saw earlier. The difference is in how you find the array's size or an array's dimension. You can use the `UBound` function to get the length of the array or the dimension. A number indicates which dimension's size is returned. Use 1 for the first dimension, 2 for the second, and so on. If the dimension is omitted, one is assumed:

```
Dim rows As Integer = Ubound(myMatrix)
Dim columns As Integer = Ubound(myMatrix,1)
```

Next, we will look at working with lists in VB.

When and how to use lists

Lists and arrays are data structures that are used in programming to store and organize collections of elements. However, there are some differences between them in terms of their properties and functionality

- **Lists:**

  - **Dynamic size**: In most programming languages, such as Python, lists are dynamic in size, meaning they can grow or shrink during program execution. You can add or remove elements from a list without specifying their size.

  - **Heterogeneous elements**: Lists can store elements of different data types. For example, a list can contain integers, strings, and objects.

- **Flexible operations**: Lists provide built-in methods for operations, such as appending elements, inserting elements at specific positions, removing elements, and concatenating lists.

- **Slower access**: Accessing elements in a list is slower than using arrays because lists use pointers to store the elements' memory addresses. As a result, accessing an element requires following the pointer to the desired memory location.

- **Higher memory overhead**: Lists generally require more memory than arrays due to their dynamic nature and additional metadata.

- **Arrays**:

  - **Fixed Size**: Arrays have a fixed size that's determined at the time of declaration and cannot be changed during runtime.

  - **Homogeneous elements**: Arrays only store elements of the same data type. For example, an integer array can only hold integer values, and a string array can only hold string values.

  - **Efficient access**: Arrays provide efficient element access since elements are stored in contiguous memory locations. Accessing an element requires calculating its memory address using an index, a simple arithmetic operation.

  - **Limited operations**: Arrays typically offer fewer built-in operations compared to lists. They usually support basic operations such as element access, modification, and iteration but may not have built-in methods for dynamic resizing or complex operations.

  - **Lower memory overhead**: Arrays generally require less memory than lists because they have a fixed size and do not require additional metadata.

To summarize the differences, lists are more flexible and versatile regarding dynamic size and heterogeneous elements. At the same time, arrays offer efficient element access and lower memory overhead but have a fixed size and homogeneous elements. The choice between lists and arrays depends on the specific requirements of your program and the operations you need to perform on the collection of elements.

In VB.NET, a list is represented by the List (Of T) class, which is part of the System.Collections. Generic namespace. The List (Of T) class provides a dynamic, resizable array-like data structure.

Here's an example of how to use lists in VB.NET:

```
Imports System.Collections.Generic
Dim nums As New List(Of Integer)()
nums.Add(10)
nums.Add(20)
nums.Add(30)
For Each num As Integer In nums
Console.WriteLine(num)
```

```
Next
Console.WriteLine("Count: " & nums.Count)
```

The preceding VB.NET code demonstrates the usage of a list to store and display a collection of integer values. Here's a breakdown of the code's functionality:

1. The code begins by importing the `System.Collections.Generic` namespace, which contains the `List(Of T)` class for creating and manipulating lists.

2. A new list called `nums` is declared and instantiated using the `New List(Of Integer)()` syntax. This list is specifically designed to store integer values.

3. Using the `Add` method, the code adds three integer elements (`10`, `20`, and `30`) to the `nums` list.

4. A `For Each` loop is utilized to iterate over each element in the `nums` list. During each iteration, the current element is assigned to the num variable, and its value is then printed to the console using `Console.WriteLine(num)`.

5. After the loop completes, the code outputs the count of elements in the `nums` list using the `Count` property. The result is printed to the console using `Console.WriteLine("Count: " & nums.Count)`.

In summary, this VB.NET code creates a list called nums and adds three integer values. It then iterates over the list and prints each element's value. Finally, it outputs the count of elements in the list.

There is also an operation to remove items from the list:

```
Nums.Remove(30)
```

VB6, **VBA**, and **VBScript** did not have native support for lists. You can instantiate COM objects in these family members as they support lists but you can also use dictionaries to get similar functionality. We will take a look at dictionaries in the next section.

Storing data in a dictionary

Dictionaries, or associative arrays or hash maps, are widely used data structures in programming. Dictionaries provide a way to store and retrieve data using key-value pairs. The key is used to store and retrieve the related data. There are also methods to return the keys as a list that can be iterated over.

In **VB.NET**, dictionaries are represented by the `Dictionary(Of TKey, TValue)` class, which is part of the `System.Collections.Generic` namespace. Dictionaries provide a data structure that stores key-value pairs, allowing efficient lookup and retrieval of values based on their associated keys.

Here's an example of how to use dictionaries in **VB.NET**:

```
Dim studGrades As New Dictionary(Of String, Integer)()
studGrades.Add("Seamus", 92)
studGrades.Add("Freya", 95)
```

```
studGrades.Add("Aspen", 89)
For Each pair As KeyValuePair(Of String, Integer) In studGrades
    Console.WriteLine(pair.Key & " Received " & pair.Value)
Next
Console.WriteLine("Students: " & studGrades.Count)
```

The preceding code snippet is written in **VB.NET** and represents a program that manages student grades using a **dictionary** data structure.

Here's a breakdown of the code:

- `Dim studGrades As New Dictionary(Of String, Integer)()`: This line declares and initializes a new dictionary instance called `studGrades`. The dictionary is expected to store key-value pairs, where the keys are of the `String` type (representing student names) and the values are of the `Integer` type (representing the grades).

- `studGrades.Add("Seamus", 92)`: This line adds a key-value pair to the `studGrades` dictionary. The key is `"Seamus"`, and the associated value is `92`.

- `studGrades.Add("Freya", 95)`: This line adds another key-value pair to the `studGrades` dictionary. The key is `"Freya"`, and the associated value is `95`.

- `studGrades.Add("Aspen", 89)`: This line adds a key-value pair to the `studGrades` dictionary. The key is `"Aspen"`, and the associated value is `89`.

- `For Each pair As KeyValuePair(Of String, Integer) In studGrades`: This line initiates a loop that iterates through each key-value pair in the `studGrades` dictionary. The loop assigns each pair to the variable pair.

- `Console.WriteLine(pair.Key & " Received " & pair.Value)`: This line prints the key (student name), followed by the " Received " string, and then the value (grade) associated with that key, to the console.

- `Next`: This keyword denotes the end of the loop initiated by the `For Each` statement.

- `Console.WriteLine("Students: " & studGrades.Count)`: This line prints the `"Students: "` string, followed by the number of items in the `studGrades` dictionary (the count of key-value pairs) to the console.

In **VBScript**, **VBA**, and **VB6**, there is no built-in dictionary data structure like the one available in **VB.NET**. You can achieve similar functionality using the `Scripting.Dictionary` object, which is provided as part of the **Windows Script Host (WSH)** object model.

To work with dictionaries in VBScript, you must create an instance of the `Scripting.Dictionary` object and then use their methods and properties to manipulate and access the data. Here's an example:

```
Dim studGrades
Set studGrades = CreateObject("Scripting.Dictionary")
```

```
studGrades.Add "Seamus", 92
studGrades.Add "Freya", 95
studGrades.Add "Aspen", 89
WScript.Echo "Seamus received " & studGrades("Seamus")
WScript.Echo "Freya received " & studGrades("Freya")
WScript.Echo "Aspen received " & studGrades("Aspen")
WScript.Echo "Students: " & studGrades.Count
```

In the previous example, we created a new instance of the `Scripting.Dictionary` object using the `CreateObject` function. Then, we used the `Add` method to add key-value pairs to the dictionary, where the keys are student names and the values are their grades.

To access the values in the dictionary, we used the key within parentheses, similar to an array index. In this case, `studGrades("Seamus")` retrieves the grade associated with the `"Seamus"` key.

The `Count` property gives the number of items (key-value pairs) in the dictionary.

Note that the `Scripting.Dictionary` object in VBScript is not as powerful or feature-rich as the dictionary in **VB.NET**. It has limited methods and does not support strongly-typed keys or values.

Summary

This chapter introduced collections in **VB** family members. We looked at working with arrays and lists and a higher-level data structure – a dictionary. These collections allow us to store multiple values in a single variable. In the next chapter, we will look at building a more extensive example project based on what we have learned.

11

Project Part II

In our second project, we will use the skills we have learned so far to build a simple student grade tool that will read student grades from a file and aggregate those grades for each VB family member. The grading tool should read a **comma-separated values (CSV)** file with a student name and grade delimited by a comma. The grading tool should utilize the dictionary and list data structures we learned about in the previous chapters. This grading tool will then show a summary per student.

In this chapter, we're going to cover the following main topics:

- Developing a grading tool in the VB.NET console
- Developing a grading tool in VB 6
- Developing a grading tool in VBA
- Developing a grading tool in VBScript
- Developing a grading tool in Classic ASP
- Developing a grading tool in ASP.NET

Technical requirements

All the example code for this chapter is available in the following GitHub repository: `https://github.com/PacktPublishing/Learn-Visual-Basics-Quick-Start-Guide-/tree/main/ProjectPartII.`

Building the project in the VB.NET console

This simple console application is written in the VB.NET programming language. This code snippet demonstrates how to read a .csv file containing student names and grades and store them in a dictionary with a list of grades. Let's break down the code:

1. The following lines import the necessary namespaces for file operations (System.IO) and working with collections (System.Collections.Generic):

    ```
    Imports System.IO
    Imports System.Collections.Generic
    ```

2. The code is enclosed within a module called MainModule; the entry point is the Main subroutine. The Main subroutine is where the execution of the program starts:

    ```
    Module MainModule
            Sub Main()
                  ' Main code goes here
            End Sub
    End Module
    ```

3. The filePath variable stores the path to the CSV file:

    ```
    Dim filePath As String = "students.csv"
    ```

4. The following line initializes a dictionary named studentGrades. The dictionary's keys are strings representing student names, and the values are lists of strings representing grades:

    ```
    Dim studentGrades As New Dictionary(Of String, List(Of
       String))()
    ```

5. The next block of code uses StreamReader to read the contents of the CSV file specified by filePath. The Using statement ensures that the reader is disposed of properly after use. The While loop continues until the end of the file is reached:

    ```
    Using reader As New StreamReader(filePath)
            While Not reader.EndOfStream
                  ' Read and process each line
            End While
        End Using
    ```

6. The following lines read a single line from the CSV file using ReadLine() and split it into fields using the Split() method. The fields are expected to be comma-separated:

    ```
    Dim line As String = reader.ReadLine()
    Dim fields As String() = line.Split(","c)
    ```

The following condition checks if the line contains precisely two fields. If it does, it proceeds to process the student's name and grade:

```
If fields.Length = 2 Then
    ' Process the student name and grade
End If
```

These lines extract the student's name and grade from the `fields` array. The `Trim()` method is used to remove any leading or trailing whitespace:

```
Dim studentName As String = fields(0).Trim()
Dim grade As String = fields(1).Trim()
```

7. This code checks whether `studentName` does not already exist as a key in the `studentGrades` dictionary. If it doesn't, a new entry is added to the dictionary with `studentName` as the key and an empty list as the initial value:

```
If Not studentGrades.ContainsKey(studentName) Then
        studentGrades.Add(studentName, New List(Of
            String)())
End If
```

8. This line adds `grade` to the list of grades associated with the `studentName` key in the `studentGrades` dictionary:

```
studentGrades(studentName).Add(grade)
```

This `For Each` loop iterates over each `studentName` instance in the Keys collection of the `studentGrades` dictionary. It allows you to process each student's name and associated grades:

```
For Each studentName As String In studentGrades.Keys
        ' Process each student name and grades
Next
```

Within the loop, this code retrieves the list of grades for the current `studentName` and joins them into a comma-separated string using `String.Join()`. It then prints the student's name and their grades to the console using `Console.WriteLine()`:

```
Dim grades As List(Of String) =
    studentGrades(studentName)
Dim gradeList As String = String.Join(", ", grades)
Console.WriteLine(studentName & ": " & gradeList).
```

9. This line waits for user input, effectively pausing the console window so that the output can be viewed before the program exits:

```
Console.ReadLine()
```

The following is the complete code for reading the CSV file, processing the data, and storing it in a dictionary with a list of grades. It then outputs the student names and their corresponding grades to the console:

```vb
Imports System.IO
Imports System.Collections.Generic

Module MainModule
    Sub Main()
        Dim filePath As String = "students.csv"
        Dim studentGrades As New Dictionary(Of String,
          List(Of String))()

        Using reader As New StreamReader(filePath)
            While Not reader.EndOfStream
                Dim line As String = reader.ReadLine()
                Dim fields As String() = line.Split(","c)

                If fields.Length = 2 Then
                    Dim studentName As String =
                      fields(0).Trim()
                    Dim grade As String = fields(1).Trim()

                    If Not studentGrades.ContainsKey
                      (studentName) Then
                        studentGrades.Add(studentName, New
                            List(Of String)())
                    End If

                    studentGrades(studentName).Add(grade)
                End If
            End While
        End Using

        For Each studentName As String In
          studentGrades.Keys
            Dim grades As List(Of String) =
              studentGrades(studentName)
            Dim gradeList As String = String.Join(", ",
              grades)
            Console.WriteLine(studentName & ": " &
              gradeList)
        Next
```

```
        Console.ReadLine()
    End Sub
End Module
```

Coding the VB.NET console grading tool

The following steps show you how to code and execute a VB.NET Windows Forms version of Hello World:

1. Start **Microsoft Visual Studio**.
2. Choose **Create New Project**.
3. Choose **Console (.NET Framework)** and click **Next**.
4. Accept **Default Project Name** and **Solution Name**.
5. Change the location if you wish your code to be in a different folder.
6. Accept the default **Framework** option.
7. Click **Create Project**.
8. Enter the code example from the preceding section.
9. To run your program, click the green triangle on the toolbar above your code.

> **Note**
> Make sure the `students.csv` file is in the same folder as the executable code; otherwise, it will throw a `System.IO.FileNotFoundException` exception.

Our next section will tackle the calculator example in VB 6.

Building the project in VB6

We will now build **VB 6** code for the grading tool program. This code reads a CSV file of students and grades and stores them in a dictionary. It uses the `Scripting.Dictionary` object to create the dictionary and a `Collection` object to store the grades for each student.

Here's the complete code, which reads the CSV file, stores the student names and grades in a dictionary, and then prints them with their corresponding grades:

```
Option Explicit
Dim studentGrades As Object
Sub ReadCSVFile()
    Dim filePath As String
    Dim fileNum As Integer
    Dim line As String
    Dim fields() As String
```

```vba
    Dim studentName As String
    Dim grade As String

    filePath = "students.csv"
    Set studentGrades = CreateObject
        ("Scripting.Dictionary")

    fileNum = FreeFile
    Open filePath For Input As fileNum

    Do Until EOF(fileNum)
        Line Input #fileNum, line
        fields = Split(line, ",")

        If UBound(fields) = 1 Then
            studentName = Trim(fields(0))
            grade = Trim(fields(1))

            If Not studentGrades.Exists(studentName) Then
                studentGrades.Add studentName, New
                    Collection
            End If

            studentGrades(studentName).Add grade
        End If
    Loop

    Close fileNum

    Dim student As Variant
    Dim gradeList as String
    For Each student In studentGrades.Keys
        studentName = student
        gradeList = GetGradeList(studentName)
        Debug.Print studentName & ": " & gradeList
    Next student
End Sub

Function GetGradeList(ByVal studentName As String) As
  String
    Dim gradeList As String
    Dim grades As Collection
    Set grades = studentGrades.Item(studentName)
```

```
    Dim grade As Variant
    For Each grade In grades
        gradeList = gradeList & grade & ", "
    Next grade

    If Len(gradeList) > 0 Then
        gradeList = Left(gradeList, Len(gradeList) - 2)
    End If

    GetGradeList = gradeList
End Function
Private Sub Form_Load()
  ReadCSVFile
End Sub
```

Let's break down the code:

1. `Option Explicit`: This statement enforces explicit declaration of variables, ensuring that all variables are declared before they are used. It helps prevent typographical errors and promotes code clarity.

2. `Dim studentGrades As Object`: This declares a variable named `studentGrades` as an `Object` type. It will be used to store the dictionary that maps student names to grades.

3. `Sub ReadCSVFile()`: This is the main subroutine and reads the CSV file and populates the dictionary with student grades.

4. `Dim filePath As String`: This declares a variable named `filePath` to store the path to the CSV file. Modify it to the actual file path.

5. `Dim fileNum As Integer`: This declares a variable named `fileNum` that stores the file number for file I/O operations.

6. `Dim line As String`: This declares a variable named `line` that stores each line read from the CSV file.

7. `Dim fields() As String`: This declares an array variable named `fields` that stores the fields that have been extracted from each line of the CSV file.

8. `Dim studentName As String`: This declares a variable named `studentName` that stores the name of each student.

9. `Dim grade As String`: This declares a variable named `grade` that stores the grade of each student.

10. `filePath = "students.csv"`: This assigns the path of the CSV file to the `filePath` variable.

11. `Set studentGrades = CreateObject("Scripting.Dictionary"):` This creates an instance of the `Scripting.Dictionary` object and assigns it to the `studentGrades` variable.

12. `fileNum = FreeFile:` This assigns a file number to the `fileNum` variable using the `FreeFile` function. It provides a unique file number for file I/O operations.

13. `Open filePath For Input As fileNum`: This opens the CSV file in input mode using the file number stored in `fileNum`.

14. `Do Until EOF(fileNum):` This starts a loop that continues until the end of the file is reached.

15. `Line Input #fileNum, line:` This reads a line from the file and assigns it to the `line` variable.

16. `fields = Split(line, ","):` This splits `line` into an array of fields using the comma delimiter.

17. `If UBound(fields) = 1 Then:` This checks if the number of fields is equal to 2. If `true`, it means there are two fields: `studentName` and `grade`.

18. `studentName = Trim(fields(0)):` This assigns the trimmed value of the first field (student name) to the `studentName` variable.

19. `grade = Trim(fields(1)):` This assigns the trimmed value of the second field (grade) to the `grade` variable.

20. `If Not studentGrades.Exists(studentName) Then:` This checks if the `studentGrades` dictionary does not already contain an entry for the current student name.

21. `studentGrades.Add studentName, New Collection:` This adds a new entry to the `studentGrades` dictionary, with the student's name as the key and a new `Collection` object as the value. The `Collection` object will store the grades for the student.

22. `studentGrades(studentName).Add grade:` This adds the current grade to the `Collection` object associated with the student name in the `studentGrades` dictionary.

23. `Loop:` This is the end of the loop. It continues reading lines from the file until the end of the file is reached.

24. `Close fileNum:` This closes the CSV file.

25. `Dim student As Variant:` This declares a variable named `student` to iterate over the keys (student names) in the `studentGrades` dictionary.

26. `For Each student In studentGrades.Keys:` This starts a loop that iterates over each student's name in the `studentGrades` dictionary.

27. `studentName = student:` This assigns the current student name to the `student Name` variable.

28. `radeList = GetGradeList(studentName):` This calls the `GetGradeList` function to retrieve the grades associated with the current student's name.

29. `Debug.Print studentName & ": " & gradeList`: This prints the student's name and the corresponding grade list to the debug output.

30. `Next student`: This is the end of the loop that iterates over the student names in the dictionary.

31. `End Sub`: This is the end of the `ReadCSVFile` subroutine.

32. `Function GetGradeList(ByVal studentName As String) As String`: This is a function that takes a student name as input and returns the corresponding grade list as a string.

33. `Dim gradeList As String`: This declares a variable named `gradeList` to store the grade list as a string.

34. `Dim grades As Collection`: This declares a variable named `grades` to store the `Collection` object associated with the student's name.

35. `Set grades = studentGrades.Item(studentName)`: This assigns the `Collection` object associated with the student's name in the `studentGrades` dictionary to the `grades` variable.

36. `Dim grade As Variant`: This declares a variable named `grade` to iterate over the grades in the `grades` `Collection` object.

37. `For Each grade In grades`: This starts a loop that iterates over each grade in the `grades` `Collection` object.

38. `gradeList = gradeList & grade & ", "`: This appends the current grade to the `gradeList` string, separated by a comma and a space.

39. `If Len(gradeList) > 0 Then`: This checks if the `gradeList` string is not empty.

40. `gradeList = Left(gradeList, Len(gradeList) - 2)`: This removes the trailing comma and space from the `gradeList` string using the `Left` function.

41. `GetGradeList = gradeList`: This assigns the `gradeList` string as the function's return value.

Coding the VB 6 grading tool

The following steps allow you to code and execute a VB 6 version of the grading tool:

1. Start **Microsoft Visual Basic 6**.

2. Choose **Standard EXE** as the project type.

3. When the designer loads, double-click inside the rectangle representing your form.

4. Enter the code example from the preceding section.

5. To run your program, click the green triangle on the toolbar above your code.

In the next section, we will tackle the grading tool example in VBA.

Building the project in VBA

What follows is a **Visual Basic for Applications** (**VBA**) application that defines a subroutine named `ReadCSVFile` that reads a CSV file containing student names and grades. Let's break down the code:

1. `Sub ReadCSVFile()`: This line starts the definition of the `ReadCSVFile` subroutine. It indicates that it doesn't return any value.

2. These lines declare several variables used in the subroutine:

    ```
    Dim fso As Object
    Dim inputFile As Object
    Dim studentGrades As Object
    Dim line As String
    Dim fields() As String
    Dim studentName As String
    Dim grade As String
    Dim grades() As String
    Dim gradeList As String
    ```

    I.   `fso`, `inputFile`, and `studentGrades` are declared as generic `Object` types and will be assigned to specific objects later.

    II. `line`, `fields()`, `studentName`, `grade`, `grades()`, and `gradeList` are declared as specific data types (`String` and `String` arrays) to store different pieces of information during the execution of the subroutine.

    These lines create objects and set them to specific objects from the `Scripting` library:

    ```
    Set fso = CreateObject("Scripting.FileSystemObject")
        Set inputFile = fso.OpenTextFile("students.csv", 1,
          False)
        Set studentGrades =
          CreateObject("Scripting.Dictionary")
    ```

    III. `fso` is set to a `FileSystemObject` object that was created using the `CreateObject` function. It provides access to the filesystem operations.

    IV. `inputFile` is set to a text file that's opened by the `OpenTextFile` method of `fso`. It opens the `students.csv` file for reading (1); `False` indicates that it should be a non-Unicode file.

    V.   `studentGrades` is set to a `Dictionary` object that was created using the `CreateObject` function. It will store the students' names as keys and their grades as values.

3. The `Do Until` loop iterates until the end of the input file is reached. It reads each line of the file using the `ReadLine` method of `inputFile` and assigns it to the `line` variable. Then, it splits `line` into fields while using a comma (`,`) as the delimiter, and stores them in the `fields` array using the `Split` function:

```
Do Until inputFile.AtEndOfStream
    line = inputFile.ReadLine
    fields = Split(line, ",")
    . . .
Loop
```

4. The `If` statement checks whether the `fields` array has a length of 2. If it does, it assumes the line contains a student's name and a grade. It assigns the first field (student name) to the `studentName` variable and the second field (grade) to the `grade` variable after trimming any leading or trailing spaces using the `Trim` function:

```
If UBound(fields) = 1 Then
    studentName = Trim(fields(0))
    grade = Trim(fields(1))
    . . .
End If
```

5. The `If-Else` statement checks whether `studentName` already exists as a key in the `studentGrades` dictionary. If it doesn't exist, it adds a new entry to the dictionary with `studentName` as the key and an array containing only `grade` as the value. If `studentName` already exists, it retrieves the existing array of grades from the dictionary, resizes it to accommodate the new grade using `ReDim Preserve`, and then assigns the new grade to the last position in the `grades` array. Finally, it updates the dictionary entry with the updated `grades` array:

```
If Not studentGrades.Exists(studentName) Then
    studentGrades.Add studentName, Array(grade)
Else
    grades = studentGrades(studentName)
    ReDim Preserve grades(UBound(grades) + 1)
    grades(UBound(grades)) = grade
    studentGrades(studentName) = grades
End If
```

6. The next line closes the input file:

```
inputFile.Close
```

7. The For Each loop iterates over each studentName in the Keys collection of the studentGrades dictionary. For each student name, it retrieves the corresponding grades array from the dictionary, joins the grades into a string separated by commas using the Join function, and then outputs the student name and grade list using Debug.Print:

```
For Each studentName In studentGrades.Keys
        grades = studentGrades(studentName)
        gradeList = Join(grades, ", ")
        Debug.Print studentName & ": " & gradeList
    Next studentName
```

Overall, this code reads the CSV file, stores the student names and grades in a dictionary, and then prints the student names along with their respective grades using Debug.Print. The loop continues until the user enters X to quit, at which point the variable sentinel is set to True, and the loop ends:

```
Option Explicit
Sub ReadCSVFile()
    Dim fso As Object
    Dim inputFile As Object
    Dim studentGrades As Object
    Dim line As String
    Dim fields() As String
    Dim studentName As String
    Dim grade As String
    Dim grades() As String
    Dim gradeList As String

    Set fso = CreateObject("Scripting.FileSystemObject")
    Set inputFile = fso.OpenTextFile("students.csv", 1,
      False)
    Set studentGrades = CreateObject
      ("Scripting.Dictionary")

    Do Until inputFile.AtEndOfStream
        line = inputFile.ReadLine
        fields = Split(line, ",")

        If Ubound(fields) = 1 Then
            studentName = Trim(fields(0))
            grade = Trim(fields(1))

            If Not studentGrades.Exists(studentName) Then
                studentGrades.Add studentName, Array(grade)
            Else
```

```
            grades = studentGrades(studentName)
            ReDim Preserve grades(Ubound(grades) + 1)
            grades(Ubound(grades)) = grade
            studentGrades(studentName) = grades
        End If
    End If
Loop

inputFile.Close

For Each studentName In studentGrades.Keys
    grades = studentGrades(studentName)
    gradeList = Join(grades, ", ")
    Debug.Print studentName & ": " & gradeList
Next studentName
End Sub
```

Coding the VBA grading tool in Excel

The following steps allow you to code and execute a VBA macro displaying `Hello World`:

1. Start **Microsoft Excel**.
2. Display the **View** menu.
3. Click the **Macros** button and choose **View Macros** from the drop-down menu.
4. Type `test` as the macro's name.
5. Click the **Create** button.
6. When the designer loads, double-click inside the rectangle representing your form.
7. Enter the code example from the preceding section.
8. Click the green triangle on the toolbar above your code to run your macro.

In the next section, we will tackle the grading tool example in VBScript.

Building the project in VBScript

The following **VBScript** program reads a CSV file containing student names and grades. It stores the data in a dictionary, where each student's name is associated with a list of their grades. Let's break down the code step by step:

1. `Set fso = CreateObject("Scripting.FileSystemObject")`: This line creates an instance of the `FileSystemObject` object, which provides access to the filesystem. It allows us to interact with files, such as opening and reading them.

2. `Set inputFile = fso.OpenTextFile("students.csv", 1, False)`: This line uses the `OpenTextFile` method of `FileSystemObject` to open the `students.csv` file in read mode (1). The `False` parameter indicates that the file should be opened as a Unicode file.

3. `Dim studentGrades`: This line declares a variable called `studentGrades` without assigning it a value. It will be used later to store the student names and grades.

4. `Set studentGrades = CreateObject("Scripting.Dictionary")`: This line creates an instance of the `Dictionary` object using the `CreateObject` function. The `Dictionary` object allows you to store key-value pairs.

5. `Do Until inputFile.AtEndOfStream`: This begins a loop that continues until the end of the input file is reached.

6. `line = inputFile.ReadLine`: This reads the next line from the input file and assigns it to the `line` variable.

7. `fields = Split(line, ",")`: This line splits the `line` variable into an array of `fields` by using the comma as the delimiter.

8. `If UBound(fields) = 1 Then`: This condition checks if the `fields` array has a length of 2. If it does, it assumes that the line contains a student's name and a grade.

9. `studentName = Trim(fields(0))` and `grade = Trim(fields(1))`: These lines assign the first field (student name) to the `studentName` variable and the second field (grade) to the `grade` variable. The `Trim` function is used to remove leading and trailing spaces from the values.

10. `If Not studentGrades.Exists(studentName) Then`: This condition checks if `studentName` already exists as a key in the `studentGrades` dictionary.

11. `studentGrades.Add studentName, Array(grade)`: If the student name doesn't exist in the dictionary, a new entry is added with `studentName` as the key and an array containing only `grade` as the value.

12. `Else`: If the student name already exists in the dictionary, this code block is executed.

13. `grades = studentGrades(studentName)`: This line retrieves the list of grades associated with `studentName` from the dictionary.

14. `ReDim Preserve grades(UBound(grades) + 1)`: This line resizes the `grades` array to accommodate the new grade. The `Preserve` keyword ensures that the existing grades are not lost during the resizing process.

15. `grades(UBound(grades)) = grade`: This line assigns the new `grade` variable to the last position in the `grades` array.

16. `studentGrades(studentName) = grades`: This updates the dictionary entry for `studentName` with the updated `grades` array.

17. `Loop`: This is the end of the loop. It continues until the end of the input file is reached.

18. `inputFile.Close`: This line closes the input file.

19. `For Each studentName In studentGrades`: This loop iterates over each student name in the `studentGrades` dictionary.

20. `grades = studentGrades(studentName)`: This line retrieves the list of grades associated with the current `studentName`.

21. `gradeList = Join(grades, ", ")`: This line joins the `grades` array elements into a string, separated by a comma and a space.

22. `WScript.Echo studentName & ": " & gradeList`: This line outputs the student name and their corresponding grade list to the console.

Overall, this code reads the CSV file, stores the student names and grades in a dictionary, and then prints the student names, along with their respective grades:

```vbscript
Set fso = CreateObject("Scripting.FileSystemObject")
Set inputFile = fso.OpenTextFile("students.csv", 1, False)

Dim studentGrades
Set studentGrades = CreateObject("Scripting.Dictionary")

Do Until inputFile.AtEndOfStream
    line = inputFile.ReadLine
    fields = Split(line, ",")

    If UBound(fields) = 1 Then
        studentName = Trim(fields(0))
        grade = Trim(fields(1))

        If Not studentGrades.Exists(studentName) Then
            studentGrades.Add studentName, Array(grade)
        Else
            grades = studentGrades(studentName)
            ReDim Preserve grades(UBound(grades) + 1)
            grades(UBound(grades)) = grade
            studentGrades(studentName) = grades
        End If
    End If
Loop

inputFile.Close

For Each studentName In studentGrades
    grades = studentGrades(studentName)
```

```
    gradeList = Join(grades, ", ")
    WScript.Echo studentName & ": " & gradeList
Next
```

Coding the VBScript grading tool with Notepad

The following steps show you how to code and execute the previous VBScript, which runs a calculator program:

1. Start Notepad.

2. Enter the code from the preceding section.

3. Save the file as `VBScriptGradesFromFile.vbs` in your root folder.

4. Run CMD to go to the Command Prompt.

5. Type `cscript VBScriptGradesFromFile.vbs` and press the *Enter* key.

> **Note**
>
> As suggested earlier, make sure you have the `students.csv` file in the same directory as you are running from.

In the next section, we will tackle the grading tool example in Classic ASP.

Building the project in Classic ASP

The following is a Classic ASP program that reads a CSV file of students and grades and then stores the data in a dictionary with a list of grades for each student. Here's a breakdown of the code:

1. The code starts with `<%` and ends with `%>`, which indicates that it is embedded within an ASP code block.

2. The `fso`, `inputFile`, `studentGrades`, `line`, `fields`, `studentName`, `grade`, `grades`, and `gradeList` variables are declared using the `Dim` statement to hold various data during processing.

3. The `Server.CreateObject` method creates an instance of the `Scripting.FileSystemObject` and `Scripting.Dictionary` objects (`fso` and `studentGrades`).

4. The `Server.MapPath` method is used to get the physical path of the `students.csv` file relative to the server's filesystem.

5. The `OpenTextFile` method of the `fso` object is called to open the `students.csv` file for reading.

6. A loop is initiated using the `Do Until` statement to read each line from the input file until the end of the file is reached (`inputFile.AtEndOfStream`).

7. Each line that's read from the file is split into fields using the `Split` function, with a comma (,) as a delimiter.

8. The `UBound` function checks whether the `fields` array has precisely two elements. If so, the student name and grade are extracted from the `fields` array and stored in the respective variables.

9. Inside the conditional block, the code checks whether the student name already exists in the `studentGrades` dictionary using the `Exists` method. If not, a new entry is added to the dictionary with the student name as the key and an array containing the grade as the value.

10. If the student name already exists in the dictionary, the existing `grades` array is retrieved using `studentName` as the key. The `ReDim Preserve` statement is used to resize the `grades` array to accommodate the new grade, and the new grade is added to the end of the array. Finally, the updated `grades` array is stored in the `studentGrades` dictionary.

11. Once all the lines have been processed, the input file is closed using the `Close` statement.

12. The code then enters a loop using the `For Each` statement to iterate through each student name in the `studentGrades` dictionary.

13. Inside the loop, the grades corresponding to the current student's name are retrieved from the dictionary. The `Join` function concatenates the grades into a comma-separated string.

14. The student name and grade list are written to the response using the `Response.Write` statement. The `<br>` tag is appended to create line breaks between each student's information.

The complete code reads the CSV file, organizes the student grades data in a dictionary, and outputs the student names with their corresponding grades in a formatted manner:

```
<%
Dim fso, inputFile, studentGrades, line, fields, studentName, grade,
grades, gradeList

Set fso = Server.CreateObject("Scripting.FileSystemObject")
Set inputFile = fso.OpenTextFile
  (Server.MapPath("students.csv"), 1, False)

Set studentGrades = Server.CreateObject
  ("Scripting.Dictionary")

Do Until inputFile.AtEndOfStream
    line = inputFile.ReadLine
    fields = Split(line, ",")

    If UBound(fields) = 1 Then
        studentName = Trim(fields(0))
        grade = Trim(fields(1))
```

```
        If Not studentGrades.Exists(studentName) Then
            studentGrades.Add studentName, Array(grade)
        Else
            grades = studentGrades(studentName)
            ReDim Preserve grades(UBound(grades) + 1)
            grades(UBound(grades)) = grade
            studentGrades(studentName) = grades
        End If
    End If
Loop

inputFile.Close

For Each studentName In studentGrades.Keys
    grades = studentGrades(studentName)
    gradeList = Join(grades, ", ")
    Response.Write studentName & ": " & gradeList & "<br>"
Next
%>
```

Coding the Classic ASP grading tool with Notepad

The following steps show you how to code and execute a VBScript that displays the calculator that we developed in this project:

1. Start Notepad.
2. Enter all the example code from the preceding section.
3. Save the file as `ClassicASPGradingTool.asp` in the root web directory – this is typically `c:\inetpub\wwwroot`.
4. Place the `students.csv` file into the same folder.
5. Open a web browser.
6. Enter `http://127.0.0.1/ClassicASPGradingTool.asp`.

In the next section, we will tackle the grading tool example in ASP.NET.

Building the project in ASP.NET

What follows is the ASP.NET code for the grading tool. This code is very similar to the code that was presented for the VB.NET console application. With this example, no ASPX code changes are required as the output is handled with the `response.write` method:

```
Imports System.IO
```

```vbnet
Public Class ASPNETGradingTool
    Inherits System.Web.UI.Page

    Protected Sub Page_Load(ByVal sender As Object, ByVal e
      As System.EventArgs) Handles Me.Load
        Dim filePath As String = "students.csv"
        Dim studentGrades As New Dictionary(Of String,
          List(Of String))()

        Using reader As New StreamReader(filePath)
            While Not reader.EndOfStream
                Dim line As String = reader.ReadLine()
                Dim fields As String() = line.Split(","c)

                If fields.Length = 2 Then
                    Dim studentName As String =
                      fields(0).Trim()
                    Dim grade As String = fields(1).Trim()

                    If Not studentGrades.ContainsKey
                      (studentName) Then
                        studentGrades.Add(studentName, New
                          List(Of String)())
                    End If

                    studentGrades(studentName).Add(grade)
                End If
            End While
        End Using

        For Each studentName As String In
          studentGrades.Keys
            Dim grades As List(Of String) =
              studentGrades(studentName)
            Dim gradeList As String = String.Join(", ",
              grades)
            Response.Write(studentName & ": " & gradeList &
              "</br>")
        Next

    End Sub

End Class
```

Coding the ASP.NET calculator with Visual Studio

The following steps will allow you to code and execute an ASP.NET web page that displays `Hello World`:

1. Start **Microsoft Visual Studio 2019**.
2. Choose **Create New Project**.
3. Choose **ASP.NET Web Application** and click **Next**.
4. Accept **Default Project Name** and **Solution Name**.
5. Change the location if you wish your code to be in a different folder.
6. Accept the default **Framework** option.
7. Click **Create Project**.
8. Choose **Web Forms**.
9. When the designer loads, navigate to **Solution Explorer** and right-click on the solution.
10. Choose **Add Item** from the pop-up menu.
11. Choose **Web Form Visual Basic**.
12. Right-click on the ASPX source and choose **View Code** from the pop-up menu.
13. Enter the VB.NET code from the code example from the preceding section that falls in the `Page_load` procedure.
14. To run your program, click the green triangle on the toolbar above your code.

Summary

In this chapter, we utilized the skills we have gained so far to build a student grading tool for each of our VB family members. This project took the skills from the project that we saw in *Chapter 7* and added the use of collections and file input. The next chapter will explore organizing our code while utilizing the **object-oriented programming (OOP)** paradigm.

Part 3:
Object-Oriented Visual Basic

In the third part of this book, we will learn to build objects that allow for more code reuse. We will learn the big three of object-oriented programming – encapsulation, inheritance, and polymorphism – and how they are used in Visual Basic.

This part has the following chapters:

12
Object-Oriented Programming

Object-oriented programming (OOP) is a programming paradigm that organizes code and data into reusable structures called objects. In our previous pieces of code, we could organize code into procedures or functions to allow for code reuse. Objects allow us to reuse code that is both executable but also stores information. In OOP, objects are instances of classes that serve as blueprints or templates for creating objects with similar properties and behaviors.

The fundamental concepts of OOP are as follows:

- **Classes**: A class is a blueprint that defines the properties (attributes) and behaviors (methods) that objects of that class will possess. It describes the structure and behavior of objects but doesn't represent any specific instance itself.

- **Objects**: An object is an instance of a class. It is created from the class blueprint and has values for attributes. Objects can interact with each other if we invoke their methods.

- **Encapsulation**: Encapsulation refers to bundling data (attributes) and methods that operate on that data into a single unit called an object. Encapsulation hides the internal details of an object and provides access to its functionality through well-defined interfaces.

- **Inheritance**: Inheritance allows classes to inherit attributes and behaviors from other classes. It enables the creation of a hierarchical structure of classes, where subclasses (derived classes) can inherit and extend the properties and methods of a superclass (base class).

- **Polymorphism**: Polymorphism means the ability of objects to take on different forms or behaviors. In the context of OOP, polymorphism allows objects of other classes to be treated as objects of a common superclass. Polymorphism allows you to write more generic code that can work with different types of objects.

- **Abstraction**: Abstraction involves the concept of creating simplified representations of complex systems. In OOP, abstraction allows you to focus on the essential features of an object while hiding the implementation details. Abstract classes and interfaces are used to define abstractions in many programming languages.

OOP provides several benefits, including code reusability, modularity, and easier maintenance. It allows for better code organization, promotes code readability, and supports the development of large-scale software systems.

Popular programming languages that support OOP include **Visual Basic** (**VB**), Java, C++, Python, C#, Ruby, and many more. Each language may have its own syntax and features for implementing OOP concepts, but the fundamental principles remain the same.

In this chapter, we're going to cover the following main topics:

- Classes and fields
- Instantiating a class
- Methods
- One-to-one has-a relationships

Classes and fields

In **VB**, classes are used to define objects with specific properties (fields) and behaviors (methods). Here's an overview of how classes and fields work in VB.

Class declaration

To define a class in VB, use the `Class` keyword followed by the class's name. Here's an example:

```
Class MyClass
    ' Class members will be defined here
End Class
```

Fields

Fields, also known as attributes or variables, are the data members of a class. They represent the properties or characteristics of an object. Fields hold data values that are associated with objects created from the class. Fields are declared within the class, typically at the beginning, and can have different access modifiers (`Public`, `Private`, `Protected`, and so on) to control their visibility and accessibility:

```
Class MyClass
    Private myField As Integer
    Public anotherField As String
End Class
```

In the preceding example, `myField` is a private field and `anotherField` is a public field.

Next, let's look at access modifiers, which we can use to control access to the fields in the class.

Access modifiers determine the accessibility of the fields within and outside the class. VB provides several access modifiers, including the following:

- `Private`: The field is only accessible within the class
- `Public`: The field is accessible from any code that can access the object
- `Protected`: The field is accessible within the class and its derived classes
- `Friend`: The field is accessible within the same assembly or project

Properties provide a controlled interface to access and modify the values of fields. They allow you to enforce validation, perform additional actions when getting or setting a value, or provide read-only or write-only access to the field. Properties are defined using the `Property` keyword in VB. Here's an example:

```
Class MyClass
    Private myField As Integer

    Public Property MyProperty As Integer
        Get
            Return myField
        End Get
        Set(value As Integer)
            myField = value
        End Set
    End Property
End Class
```

In the preceding example, `MyProperty` is a property that provides access to the `myField` field.

Fields and properties in VB enable you to store and manipulate data within objects of a class. They are an integral part of defining the state and behavior of objects and play a crucial role in OOP.

All the VB family members support class creation except VBScript. VBScript can use classes defined in **ActiveX** objects, as seen with the filesystem and collections.

Next, we will look at working with methods in VB.

Methods

In VB, methods define the behavior or actions that objects can perform. Methods are described within classes and are responsible for performing specific tasks or calculations.

Method declaration

To define a method in VB, use the `Sub` or `Function` keyword, followed by the method's name and any parameters it accepts. Here's an example:

```
Public Sub MyMethod(parameter1 As Integer, parameter2 As
   String)
     ' Method body
End Sub
```

In the preceding example, `MyMethod` is a public method that accepts an integer parameter named `parameter1` and a string parameter named `parameter2`.

Next, let's drill into the types of methods we can have in our classes:

- **Sub procedures**: Sub procedures are methods that do not return a value. They are typically used for performing actions or tasks without producing a result. Here's an example:

  ```
  Public Sub DisplayMessage(message As String)
      Console.WriteLine(message)
  End Sub
  ```

  In the preceding example, `DisplayMessage` is a public method that accepts a string parameter named `message`. It displays the message on the console.

- **Function procedures**: Function procedures are methods that return a value. They are used when you need to perform calculations or retrieve data from a method. Here's an example:

  ```
  Public Function AddNumbers(a As Integer, b As Integer)
    As Integer
       Return a + b
  End Function
  ```

  In the preceding example, `AddNumbers` is a public method that accepts two `Integer` parameters, a and b, and returns their sum as an integer value.

- **Access modifiers**: Like fields, methods can have different access modifiers to control their visibility and accessibility. Common access modifiers in VB include `Public`, `Private`, `Protected`, and `Friend`.

- **Method overloading**: VB supports method overloading, which allows you to define multiple methods with the same name but different parameters. The compiler determines which method to call based on the arguments provided during the method invocation. Here's an example of method overloading:

  ```
  Public Sub DoSomething()
      ' Method implementation without parameters
  End Sub
  ```

```
Public Sub DoSomething(parameter As Integer)
    ' Method implementation with an integer parameter
End Sub

Public Sub DoSomething(parameter As String)
    ' Method implementation with a string parameter
End Sub
```

In the preceding example, DoSomething is overloaded with three different versions based on the presence and type of parameters.

Methods in VB enable you to encapsulate functionality within classes and provide a way to perform actions or calculations. They allow for code reuse, modularity, and better logic organization within your programs.

Next, we will look at instantiating a class in VB.

Instantiating a class

In VB, you can instantiate classes by creating objects from those classes. The process involves using the New keyword, followed by the class name and any necessary arguments. Here's an example of how to instantiate a class in VB:

```
Dim myObject As New ClassName()
```

In the preceding example, ClassName represents the name of the class you want to instantiate, and myObject is the variable that will hold the created object.

If the class has a parameterized constructor, you can pass arguments when instantiating the class:

```
Dim myObject As New ClassName(arg1, arg2)
```

In this case, arg1 and arg2 are the arguments required by the constructor of ClassName.

Once you have instantiated a class, you can access its properties, methods, and other members through the object variable. Here's an example:

```
myObject.PropertyName = value ' Set the value of a property
myObject.MethodName() ' Call a method
```

It's important to note that the process of instantiating a class creates a new instance of the class, which means that the object and its associated data are stored in memory separately from other instances of the same class.

Additionally, you can create multiple instances of a class, each with its own set of data. This allows you to work with different objects based on the same class blueprint.

Remember to choose meaningful names for your class and object variables to ensure clarity and maintainability in your code.

Next, we will look at representing one-to-one has-a relationships in VB.

One-to-one has-a relationships

In VB, a one-to-one has-a relationship between classes refers to a situation where one class contains an instance of another class as a field or property. This relationship represents an association between the two classes, where the first class directly references an instance of the second class. Here's an example of a one-to-one has-a relationship in VB:

```
Public Class Car
    Public Property Engine As Engine
    ' Other car-related properties and methods
End Class
Public Class Engine
    ' Engine-related properties and methods
End Class
```

In the preceding example, we have two classes: `Car` and `Engine`. The `Car` class has a property called `Engine` of the `Engine` type and represents the car's engine. This relationship indicates that a car has a single engine.

To use this relationship, you would typically create instances of the classes and set the appropriate values:

```
Dim myCar As New Car()
myCar.Engine = New Engine()
```

In the preceding code snippet, we create a new `Car` object named `myCar` and assign a new `Engine` object to its `Engine` property.

With this one-to-one relationship, the `Car` class can access the properties and methods of the associated `Engine` class through its `Engine` property:

```
myCar.Engine.Start()
```

It's important to note that a one-to-one has-a relationship in VB can also be implemented using fields instead of properties, depending on your specific needs and design preferences.

One-to-one relationships can be helpful when different classes need to collaborate and share information or functionality. They allow for modular and extensible code by separating the concerns into other classes while maintaining a direct association between them.

Summary

This chapter introduced OOP in VB. We looked at defining classes and fields in VB. This chapter will be the basis for the following three chapters, where we'll drill into specific object-oriented software development techniques. The next chapter will look at building taxonomies while utilizing inheritance in VB programming.

13
Inheritance

Inheritance is a fundamental concept in **object-oriented programming** (**OOP**) that allows a child class to inherit properties and behaviors from another class (parent or base class). Inheritance enables code reuse, promotes a hierarchical organization of classes, and facilitates the creation of a more organized and maintainable code base.

Inheritance follows the *is-a* relationship, meaning a child class is a specialized version of the parent class. It can access all the parent class's public and protected members (variables and methods) but not private ones. The child class can also override the methods of the parent class to provide its specific implementation while keeping the rest of the inherited behaviors intact.

In this chapter, we're going to cover the following main topics:

- Inheritance in **VB.NET**
- Inheritance differences in family members
- Exception handling
- Is-a relationships

Inheritance in VB.NET

In **VB.NET** inheritance, a class can inherit properties, methods, and events from another class, referred to as the base class or parent class. The class that inherits from the base class is called a child class. Inheritance allows you to create a hierarchy of classes, promoting code reuse and a more organized code structure.

To demonstrate inheritance in VB.NET, let's create an example with a base class, `Animal`, and two derived classes, `Dog` and `Cat`:

```
Public Class Animal
    Public Property Name As String
Talk
```

```
        Public Overridable Function Talk() As String
            Return "animal noise"
        End Function
    End Class
```

The Dog class will inherit from the Animal class:

```
Public Class Dog Inherits Animal
    Public Property Breed As String
    Public Overrides Function Talk() As String
        Return "Bark!"
    End Function
End Class
```

The Cat class will also inherit from the Animal class:

```
Public Class Cat Inherits Animal
    Public Property Color As String
    Public Overrides Function Talk() As String
        Return "Meow!"
    End Function
End Class
```

We can use these classes in code:

```
Module Module1
    Sub Main()
        Dim myDog As New Dog()
        myDog.Name = "Maisie"
        myDog.Breed = "Dalmatian"
        Console.WriteLine(myDog.Name + " - " + myDog.Breed
            + " - " + myDog.Talk())
        Dim myCat As New Cat()
        myCat.Name = "Woody"
        myCat.Color = "Black"
        Console.WriteLine(myCat.Name + " - " + myCat.Color
            + " - " + myCat.Talk())
    End Sub
End Module
```

In the preceding code block, we created a base class called Animal with a Name property and a Talk method. The Talk method is marked as Overridable, which means it can be overridden in the derived classes.

Then, we created two derived classes, Dog and Cat, each inheriting from Animal. In the derived classes, we added specific properties (Breed for Dog and Color for Cat) and overrode the Talk method to provide their implementations of the sound they make.

Using inheritance, we can define typical behavior in the base class and extend it in the derived classes, promoting code reuse and maintaining a clear hierarchy of classes.

Next, we will explore the difference in inheritance in **VB** family members.

Inheritance differences in family members

Inheritance in VB.NET and VB6/VBA share the same core concept of allowing classes to inherit properties and behaviors from other classes. However, there are significant differences in how inheritance is implemented and supported between the two versions of VB. Let's look at these differences.

VB6/VBA

- **Limited support**: VB6 supports basic OOP, including inheritance. However, it lacks some essential features that are found in modern object-oriented languages.

- **Class modules**: In VB6, you create classes using class modules. Each class can have properties, methods, and events.

- **Inheritance keyword**: VB6 uses the `Inherits` keyword to establish an inheritance relationship between derived and base classes. However, VB6 does not support the inheritance of implementation details as it lacks the concept of access modifiers such as `public`, `protected`, or `private`.

- **Interface-based inheritance**: VB6 supports interface-based inheritance, allowing a class to implement one or more interfaces to provide a set of specific functionalities. However, VB6 does not support multiple inheritance of classes, meaning a class can only inherit from a single base class.

VB.NET

- **Full object-oriented support**: VB.NET is a fully OOP language, and it is part of the **.NET** framework, which provides extensive support for OOP principles, including inheritance.

- **Classes and access modifiers**: In VB.NET, classes are created using the `Class` keyword. You can define properties, methods, events, and access modifiers (`public`, `private`, `protected`, and so on) to control the visibility of members within the class hierarchy.

- **Inheritance keywords**: VB.NET uses the `Inherits` keyword to denote inheritance relationships between classes, similar to VB6. However, it also uses the `Overrides` keyword to indicate that a method in the derived class overrides one in the base class.

- **Multiple inheritance through interfaces**: VB.NET supports multiple inheritance through interfaces, allowing a class to implement multiple interfaces to provide various functionalities. However, multiple inheritance of classes is not directly supported to avoid issues related to the "diamond problem."

- **Abstract and sealed classes**: VB.NET allows you to define abstract classes (using the MustInherit keyword) that cannot be instantiated but can serve as base classes for other classes. Additionally, you can create sealed classes (using the NotInheritable keyword) to prevent further inheritance from that class.

VB.NET offers more robust support for OOP and inheritance than VB6. It introduces modern OOP concepts, access modifiers, interface-based inheritance, and more advanced language features.

Next, we will look at handling errors, which requires an understanding of inheritance.

Exception handling

In VB, exception handling is primarily accomplished using structured error-handling constructs, such as Try...Catch...Finally blocks, rather than inheritance. However, it's worth noting that exception handling in VB does involve classes and inheritance but in a different manner.

VB uses the Exception class, which is a base class for all exceptions in the .NET Framework, to handle errors and exceptions. The Exception class has several derived classes representing specific types of exceptions, such as SystemException and ApplicationException. When an exception occurs, VB throws an object of one of these derived exception classes, which you can then catch and handle using Try...Catch...Finally blocks.

Here's an example demonstrating exception handling in VB using Try...Catch...Finally:

```
Sub Main()
    Try
        Dim n1 As Integer = 20
        Dim n2 As Integer = 0
        Dim result As Integer = n1 \ n2
        Console.WriteLine("Result: " & result)

    Catch ex As DivideByZeroException
        Console.WriteLine("Error: " & ex.Message)

    Catch ex As Exception
        Console.WriteLine("Error: " & ex.Message)

    Finally
        Console.WriteLine("Finally block executed.")
    End Try

    ' Rest of the code continues after exception handling.
End Sub
```

In this example, we attempted to divide an integer by zero, which resulted in `DivideByZeroException`. The `Try...Catch...Finally` block allows us to handle this exception gracefully, providing custom error messages or performing cleanup operations in the `Finally` block.

While inheritance isn't directly involved in the exception-handling process, the `Exception` class and its derived classes represent a hierarchy of exceptions that can be used to catch specific types of errors and provide appropriate error handling based on the particular exception type. In this sense, inheritance plays a role in the design of the exception hierarchy.

Next, we will look at is-a relationships and taxonomies.

Is-a relationships

The *is-a* relationship is a fundamental concept in OOP that describes the inheritance relationship between classes. It is also known as the "inheritance relationship" or "subclass-superclass relationship." The is-a relationship signifies that one class is a specialized version of another, emphasizing a hierarchical classification of objects.

When we say Class A is-a Class B, this means that Class A is a specific type of Class B. Class A inherits the properties, methods, and behavior of Class B, and it can also have additional properties and methods specific to its nature.

In the example with cats and dogs we built earlier in this chapter, we can say the following:

- `Dog` is-a `Animal`
- `Cat` is-a `Animal`

These is-a statements imply that `Dog` and `Cat` are specialized types of `Animal`, inheriting the common properties and methods defined in the `Animal` class.

The is-a relationship enables code reuse and promotes a hierarchical organization of classes, leading to more organized and maintainable code. It also allows you to model real-world relationships and hierarchies in your software, making it more intuitive and easier to understand.

In OOP, the is-a relationship is often used as the basis to build class hierarchies, where more specific classes inherit from more general classes, leading to a more efficient and structured design.

Summary

This chapter introduced inheritance in VB. First, we looked at defining taxonomies in VB that allow for enhanced code reuse. Then, we drilled into the concept of is-a relationships, where children classes are instances of the parent class. The next chapter will utilize these taxonomies to implement polymorphism in VB programming.

14

Polymorphism

Polymorphism is a fundamental concept in programming and is often associated with **object-oriented programming** (**OOP**). It allows objects of different classes to be used as objects of a common superclass or interface, providing a consistent interface to interact with these objects regardless of their specific types.

In this chapter, we're going to cover the following main topics:

- Polymorphism
- Abstract classes
- Sealed classes

Polymorphism

Polymorphism allows a programmer to specialize functionality based on a context; the programming language will use the proper version. There are two main types of polymorphism:

- **Compile-time (static) polymorphism**: Also known as method overloading or compile-time method dispatch, this type of polymorphism occurs when different methods or functions have the same name but different parameter lists. The decision on which method to call is made at compile time based on the number and types of arguments provided. This allows you to have multiple methods with the same name but different behaviors based on the input parameters.

- **Runtime (dynamic) polymorphism**: Also known as method overriding or runtime method dispatch, this type of polymorphism occurs when a subclass provides a specific implementation of a method already defined in its superclass (or an interface). At runtime, the method call is resolved to the subclass implementation if the object referenced is an instance of the subclass. This allows you to use a common interface to invoke different behaviors based on the actual type of the object.

In **Visual Basic (VB)**, as in other OOP languages, polymorphism is achieved through inheritance and method overriding. VB supports runtime polymorphism, where the objects of derived classes can be treated as objects of their base class, and the appropriate method implementation is invoked based on the actual type of the object at runtime.

To demonstrate polymorphism in VB, we can create a simple example with a base class and its derived classes:

```vb
' Base class
Public Class Shape
    Public Overridable Sub Draw()
        Console.WriteLine("Drawing a shape")
    End Sub
End Class

' Derived classes
Public Class Circle
    Inherits Shape

    Public Overrides Sub Draw()
        Console.WriteLine("Drawing a circle")
    End Sub
End Class

Public Class Square
    Inherits Shape

    Public Overrides Sub Draw()
        Console.WriteLine("Drawing a square")
    End Sub
End Class
```

In this example, we have a base class called `Shape`, with a virtual method called `Draw`, and two derived classes called `Circle` and `Square`, both of which override the `Draw` method.

Next, let's use these classes to demonstrate polymorphism:

```vb
Module MainModule
    Sub Main()
        Dim shape1 As Shape = New Shape()
        Dim shape2 As Shape = New Circle()
        Dim shape3 As Shape = New Square()

        shape1.Draw() ' Output: Drawing a shape
        shape2.Draw() ' Output: Drawing a circle
```

```
        shape3.Draw() ' Output: Drawing a square
    End Sub
End Module
```

In this example, we created objects of different types (base class and derived classes) and stored them in variables of the base class type (Shape). When we call the Draw method on each object, the appropriate implementation is invoked based on the actual type of the object.

In VB, the Overridable keyword is used to declare a method in the base class, which can be overridden in derived classes. The Overrides keyword is used in the derived classes to indicate that they provide a specific implementation of the method from the base class. This way, VB enables runtime polymorphism, allowing for more flexible and extensible code.

Next, we'll look at building abstract classes, which require a specialization class to allow instantiation.

Abstract classes

In **VB**, abstract classes cannot be instantiated directly but serve as base classes for other classes. They are used to define a standard interface and behavior that derived classes must implement. Abstract classes can contain abstract (must be overridden) and non-abstract (can have default implementations) methods, properties, and fields.

To create an abstract class in VB, you can use the MustInherit keyword:

```
Public MustInherit Class Shape
    Public MustOverride Sub Draw()

    Public Sub GetArea()
        Console.WriteLine("Calculating area...")
    End Sub
End Class
```

In this example, we defined an abstract class called Shape, with an abstract method called Draw, and a non-abstract method called GetArea. The Draw method is marked with the MustOverride keyword, indicating that any derived class must provide its implementation for the Draw method.

Next, let's create a derived class that inherits from the abstract class and implements the Draw method:

```
Public Class Circle
    Inherits Shape

    Public Overrides Sub Draw()
        Console.WriteLine("Drawing a circle")
    End Sub
End Class
```

In this example, the `Circle` class inherits from the `Shape` abstract class and provides an implementation for the `Draw` method.

Since abstract classes cannot be directly instantiated, we can use them through their derived classes:

```
Module MainModule
    Sub Main()
        Dim circle As New Circle()
        circle.Draw() ' Output: Drawing a circle
        circle.GetArea() ' Output: Calculating area...
    End Sub
End Module
```

In this `Main` method, we created an instance of the `Circle` class and called its `Draw` method, which is defined in the `Shape` abstract class. We can also call the non-abstract `GetArea` method provided by the abstract class.

Abstract classes are useful when you want to define a common structure and behavior across multiple related classes while enforcing that certain methods must be implemented in the derived classes. They provide a level of abstraction and help organize code effectively in OOP.

Next, we'll look at ways to ensure that a class cannot be extended through inheritance.

Sealed classes

In VB, there is no direct equivalent keyword to *sealed*, which exists in some other programming languages, such as **C#**. In C#, the `sealed` keyword is used to prevent a class from being inherited (that is, it cannot serve as a base class). The sealed concept is important if the programmer has code that should never be specialized. An example could be an authorization class that validates a user and you do not want changes to the authorization method. However, in VB, all classes are inheritable by default, meaning that any class can be used as a base class unless it is specifically designed as an abstract class (using the `MustInherit` keyword, as explained in the previous section). To prevent a class from being inherited in VB, you can use the `NotInheritable` keyword when defining the class. The `NotInheritable` keyword is the equivalent of the `sealed` keyword in C#.

Here's an example of a `NotInheritable` class in VB:

```
NotInheritable Public Class Singleton
    Private Shared instance As Singleton

    Private Sub New()
        ' Private constructor to prevent direct
            instantiation.
    End Sub
```

```
    Public Shared Function GetInstance() As Singleton
        If instance Is Nothing Then
            instance = New Singleton()
        End If
        Return instance
    End Function


    ' Other methods and properties...
 End Class
```

In this example, the `Singleton` class is marked as `NotInheritable`, meaning no other class can inherit from it. The `Singleton` class is often used when you want to create a class with only one instance (the Singleton pattern) and ensure that no one can create additional instances by inheriting from it.

In VB, you can use the `NotInheritable` keyword to prevent a class from being inherited, similar to the `sealed` keyword in C#.

Summary

This chapter introduced polymorphism in VB. We looked at defining classes that do not have a complete definition but can be extended down in the inheritance tree. We also looked at ways to implement functionality to enforce that a class cannot be extended with inheritance. The next chapter will look at utilizing interfaces in VB programming to allow for multiple inheritance.

15
Interfaces

In **object-oriented programming** (**OOP**), interfaces are a fundamental concept that allows you to define a contract or a set of method signatures that classes must implement. An interface serves as a blueprint for implementing classes, ensuring that specific methods are available in those classes. Interfaces enable polymorphism and provide a way to achieve abstraction and separation of concerns in OOP.

In this chapter, we're going to cover the following main topics:

- Interfaces
- Developing interfaces
- Standard interfaces

Interfaces

The interface contract has been a tremendous tool for team programming and software extensibility. The interface contract specifies the method name, number of parameters, and data type of the parameters. Here are the key characteristics and benefits of using interfaces in OOP:

- **Abstraction**: Interfaces help in achieving abstraction by defining a clear and standardized set of method signatures without specifying their implementation details. This allows you to hide the internal complexity of classes and focus on the behavior they expose.

- **Separation of concerns**: By using interfaces, you can separate the interface of a class from its implementation. This enables you to change or extend the behavior of a class without it affecting the client code that uses the interface.

- **Polymorphism**: Interfaces play a crucial role in achieving polymorphism. When a class implements an interface, it can be treated as an instance of that interface, allowing objects of different classes to be used interchangeably if they implement the same interface.

- **Multiple inheritance (interface inheritance)**: In languages that support multiple inheritance, a class can implement multiple interfaces. This allows a class to inherit behavior from multiple sources, promoting code reuse and flexibility.

- **Contractual obligation**: When a class implements an interface, it must provide concrete implementations for all the methods declared in the interface. This contractual obligation ensures consistency and ensures that all implementing classes adhere to the specified interface.

- **Interface segregation principle (ISP)**: Interfaces support the ISP, which states that a class should not be forced to implement interfaces that it does not use. By defining specific interfaces with well-defined responsibilities, you can adhere to the ISP and promote cleaner and more maintainable code.

Interfaces are powerful tools in OOP, helping build flexible and maintainable software by promoting a standard way of interacting with classes and enabling polymorphic behavior across different classes.

Next, we'll look at developing new interfaces.

Developing interfaces

In **VB**, interfaces work similarly to other OOP languages. An interface in VB defines a contract that specifies a set of method signatures that classes must implement. By implementing an interface, a class agrees to provide concrete implementations for all the methods declared in the interface.

Next, we will see how to define and use interfaces in VB.

Interface declaration

To create an interface in VB, you can use the `Interface` keyword, followed by the interface's name:

```
Public Interface IShape
    Sub Draw()
    Function GetArea() As Double
End Interface
```

In this example, we defined an `IShape` interface with two methods: `Draw` and `GetArea`. These methods are declared but not implemented in the interface; their implementation will be provided by the classes that implement this interface.

Implementing an interface

To implement an interface, a class must use the `Implements` keyword, followed by the interface's name. The class then provides concrete implementations for all the methods declared in the interface:

```
Public Class Circle
    Implements IShape
```

```
        Private radius As Double
        Public Sub New(ByVal radius As Double)
            Me.radius = radius
        End Sub

        Public Sub Draw() Implements IShape.Draw
            Console.WriteLine("Drawing a circle")
        End Sub

        Public Function GetArea() As Double Implements
          IShape.GetArea
            Return Math.PI * radius * radius
        End Function
    End Class
```

In this example, the `Circle` class implements the `IShape` interface by providing concrete implementations for the `Draw` and `GetArea` methods declared in the interface.

Using interface polymorphism

Once a class implements an interface, objects of that class can be treated as instances of the interface. This allows for polymorphism, where different classes that implement the same interface can be used interchangeably:

```
Sub Main()
    Dim circle As New Circle(5)
    DrawShape(circle)
End Sub

Sub DrawShape(ByVal shape As IShape)
    shape.Draw()
End Sub
```

In this example, the `DrawShape` method takes an `IShape` interface as a parameter. The `circle` object of the `Circle` class is passed to this method, and the `Draw` method is called on it, providing the polymorphic behavior.

In VB, interfaces play a crucial role in achieving abstraction, separation of concerns, and polymorphism in OOP, just like in other languages. They help create more flexible and maintainable code by defining contracts that classes must adhere to, without specifying the implementation details.

Next, we will look at some standard interfaces provided by **VB.NET**.

Standard interfaces

In VB, standard interfaces are predefined interfaces that are part of the **.NET** framework or other external libraries. These interfaces are commonly used in VB and provide standardized behavior for various functionalities. By adhering to these standard interfaces, your classes can integrate seamlessly with other parts of the .NET ecosystem and follow best practices.

Here are some of the standard interfaces in VB.NET:

- `IEnumerable` and `IEnumerator`: These interfaces are part of the `System.Collections` namespace and enable iteration over collections such as arrays, lists, and other data structures.

- `IDisposable`: The `IDisposable` interface, which is part of the `System` namespace, is used to implement proper resource cleanup for objects that use unmanaged resources, such as file handles or database connections.

- `IComparable` and `IComparer`: These interfaces, which are also part of the `System.Collections` namespace, allow for custom sorting of objects and collections.

- `INotifyPropertyChanged` and `INotifyCollectionChanged`: These interfaces, which can be found in the `System.ComponentModel` namespace, are used to provide notifications when properties or collections change their values, which is helpful for data binding and user interface updates.

- `ICloneable`: The `ICloneable` interface in the `System` namespace provides a method for creating a shallow copy of an object.

- `IDictionary` and `IList`: These interfaces, which are part of the `System.Collections` namespace, define the basic operations for working with dictionaries (key-value pairs) and lists (indexed collections).

- `IEquatable`: The `IEquatable` interface, which is part of the `System` namespace, provides a method to compare objects for equality. It's used in conjunction with the `Equals` method.

- `IAsyncResult`: The `IAsyncResult` interface, which can be found in the `System` namespace, is used for asynchronous programming to represent the status and results of asynchronous operations.

- `IFormattable`: The `IFormattable` interface in the `System` namespace allows objects to control how they are formatted when they're converted into a string representation.

- `IConvertible`: The `IConvertible` interface, also part of the `System` namespace, allows objects to be converted into other data types.

These are just a few examples of the standard interfaces available in VB/.NET. By implementing these interfaces in your classes, you can follow established patterns and conventions, making your code more consistent and easier to integrate with other parts of the .NET ecosystem.

To demonstrate the use of a standard interface in VB.NET, we will drill into the `IComparable` interface. To use the `IComparable` interface, you need to implement it in your class. Here's an example of how you can do this:

```vbnet
Public Class Person
    Implements IComparable

    Public Property Name As String
    Public Property Age As Integer

    Public Sub New(ByVal name As String, ByVal age As
        Integer)
        Me.Name = name
        Me.Age = age
    End Sub

    Public Function CompareTo(ByVal obj As Object) As
        Integer Implements IComparable.CompareTo
        If TypeOf obj Is Person Then
            Dim otherPerson As Person = DirectCast(obj,
                Person)

            ' Compare based on age
            Return Me.Age.CompareTo(otherPerson.Age)
        End If

        Throw New ArgumentException("Object is not a
            Person")
    End Function
End Class
```

In this example, we have a `Person` class that implements the `IComparable` interface. The `CompareTo` method compares two `Person` objects based on their age. If you want to sort a collection of `Person` objects, you can use sorting algorithms that rely on the `IComparable` interface, such as `Array.Sort` or `List.Sort`.

Here's an example of how you can use the `IComparable` interface to sort a list of `Person` objects:

```vbnet
Dim people As New List(Of Person)()
people.Add(New Person("Freya", 16))
people.Add(New Person("Seamus", 15))
people.Add(New Person("Kirsten", 52))

people.Sort()
```

```
For Each p As Person In people
    Console.WriteLine($"{p.Name} - {p.Age}")
Next
```

This will output the sorted list of people based on their age.

Summary

This chapter introduced interfaces in VB. We looked at defining new interfaces in VB and utilizing standard interfaces provided in the VB language. The next chapter will look at utilizing the OOP techniques we have learned about over the last several chapters in a VB programming project.

16

Project Part III

In our third project, we will use the skills we learned in *Part 3* of this book to build a simple student-grade tool to read student grades from a file. Depending on the assessment category, this project version will support different data in each row. We will utilize inheritance so that each grade is a generalized object we can aggregate and has a specialized version for each assessment type. The grading tool should read a **comma-separated values** (**CSV**) file with a student name and grade delimited by a comma. The third column of each row will be the assessment type. We will support three assessment types: Test, Quiz, and HW. For each Test grade, the fourth field will be the time spent taking the test. We will get the number of takes for each quiz in the fourth field. For HW, we will store the submitted date. The grading tool should utilize the dictionary we learned about in recent chapters. The Object-Oriented Grading Tool will then show a summary per student by grade category. We will limit this project's scope to **VB.NET** and **ASP.NET** as the older **Visual Basic** family members do not support inheritance.

In this chapter, we're going to cover the following main topics:

- Developing an Object-Oriented Grading Tool in the VB.NET console
- Developing an Object-Oriented Grading Tool in ASP.NET

Technical requirements

All the example code for this chapter is available in the following GitHub repository: `https://github.com/PacktPublishing/Learn-Visual-Basics-Quick-Start-Guide-/tree/main/ProjectPartIII`

Building the project in the VB.NET console

This simple console application is written in the **VB.NET** programming language. We will start by exploring the different classes we must develop for the project. The following code defines a **VB.NET** class named `clsGrade` that encapsulates a concept related to grading or scoring. It provides a private field, `piScore`, to store the numerical score, and a public property, `Score`, that allows getting and setting the value of the score.

Here's a breakdown of the code that follows:

- `Public Class clsGrade`: This line declares a public class named `clsGrade`.

- `Private piScore As Integer`: This line declares a private field named `piScore` of type `Integer`, which is used to store the numeric score.

- `Public Property Score As Integer`: This line defines a public property named `Score`. Properties allow access to or modify private fields while encapsulating the implementation details.

- `Get`: This block defines the getter for the `Score` property. When the property is accessed, this block is executed and returns the value of the `piScore` private field.

- `Set(value As Integer)`: This block defines the setter for the `Score` property. When the property is assigned a value, this block is executed and assigns the provided value to the `piScore` private field.

The use of properties in this code allows you to interact with the `piScore` field in a controlled manner. The property provides a level of abstraction, allowing you to get and set the score without directly accessing the underlying field. This can be useful for maintaining data integrity, applying validation, or performing additional actions when the value is accessed or modified. Here is the code sample:

```
Public Class clsGrade
    Private piScore As Integer

    Public Property Score As Integer
        Get
            Return piScore
        End Get
        Set(value As Integer)
            piScore = value
        End Set
    End Property
End Class
```

The next section of code defines a **VB.NET** class named `clsTestGrade` that extends or inherits from another class named `clsGrade`. The `clsTestGrade` class represents a test grade, and it adds functionality specific to the duration of the test in addition to the score.

Here's a breakdown of the code, which follows this breakdown and explanation:

- `Public Class clsTestGrade`: This line declares a class named `clsTestGrade`, intended to represent a test grade.

- `Inherits clsGrade`: This line indicates that the `clsTestGrade` class inherits from another class called `clsGrade`. This means that `clsTestGrade` inherits the members (fields, properties, methods, and so on) and behavior of the `clsGrade` class.

- `Private piDuration As Integer`: This line declares a private field named `piDuration` of type `Integer`, which stores the test duration.

- `Public Sub New(piScore As Integer, piDuration As Integer)`: This is a constructor for the `clsTestGrade` class. It takes two parameters: `piScore` and `piDuration`. When an instance of `clsTestGrade` is created, this constructor is called to initialize its properties.

- `Me.Score = piScore`: This line sets the `Score` property of the `clsTestGrade` instance to the value of the `piScore` parameter. The `Score` property is inherited from the `clsGrade` class.

- `Me.piDuration = piDuration`: This line sets the `piDuration` field of the `clsTestGrade` instance to the value of the `piDuration` parameter.

- `Public Property Duration As Integer`: This line defines a public property named `Duration`. This property allows getting and setting the duration of the test.

- `Get`: This block defines the getter for the `Duration` property. When the property is accessed, this block is executed and returns the value of the `piDuration` private field.

- `Set(value As Integer)`: This block defines the setter for the `Duration` property. When the property is assigned a value, this block is executed and assigns the provided value to the `piDuration` private field.

This class is designed to work with test grades and includes an additional attribute for test duration. The inheritance from the `clsGrade` class provides common properties or methods related to grades, and `clsTestGrade` extends it to include specific behavior related to test grading and duration. Here is the code:

```
Public Class clsTestGrade
    Inherits clsGrade
    Private piDuration As Integer

    Public Sub New(piScore As Integer, piDuration As
      Integer)
        Me.Score = piScore
        Me.piDuration = piDuration
    End Sub

    Public Property Duration As Integer
        Get
            Return piDuration
        End Get
        Set(value As Integer)
            piDuration = value
```

```
        End Set
    End Property
End Class
```

In the next code section, we define a **VB.NET** class named `clsQuizGrade` that extends or inherits from another class named `clsGrade`. The `clsQuizGrade` class represents a quiz grade and includes information about the number of attempts made for the quiz and the score.

Here's a breakdown of the code, which follows this breakdown and explanation:

- `Public Class clsQuizGrade`: This line declares a class named `clsQuizGrade` intended to represent a quiz grade.

- `Inherits clsGrade`: This line indicates that the `clsQuizGrade` class inherits from another class called `clsGrade`. The `clsGrade` class contains common properties or methods related to grading, and `clsQuizGrade` extends it to include specific behavior related to quiz grading.

- `Private piAttempts As Integer`: This line declares a private field named `piAttempts` of type `Integer`, which stores the number of attempts made for the quiz.

- `Public Sub New(piScore As Integer, piAttempts As Integer)`: This is a constructor for the `clsQuizGrade` class. It takes two parameters: `piScore` and `piAttempts`. When an instance of `clsQuizGrade` is created, this constructor is called to initialize its properties.

- `Me.Score = piScore`: This line sets the `Score` property of the `clsQuizGrade` instance to the value of the `piScore` parameter. The `Score` property is inherited from the `clsGrade` class.

- `Me.piAttempts = piAttempts`: This line sets the `piAttempts` field of the `clsQuizGrade` instance to the value of the `piAttempts` parameter.

- `Public Property Attempts As Integer`: This line defines a public property named `Attempts`. This property allows getting and setting the number of attempts made for the quiz.

- `Get`: This block defines the getter for the `Attempts` property. When the property is accessed, this block is executed and returns the value of the `piAttempts` private field.

- `Set(value As Integer)`: This block defines the setter for the `Attempts` property. When the property is assigned a value, this block is executed and assigns the provided value to the `piAttempts` private field.

This class is designed to work with quiz grades and includes an additional attribute for the number of attempts made for the quiz. The inheritance from the `clsGrade` class provides standard grading-related functionality, and `clsQuizGrade` extends it to include quiz-specific behavior:

```
Public Class clsQuizGrade
    Inherits clsGrade
    Private piAttempts As Integer

    Public Sub New(piScore As Integer, piAttempts As
      Integer)
        Me.Score = piScore
        Me.piAttempts = piAttempts
    End Sub

    Public Property Attempts As Integer
        Get
            Return piAttempts
        End Get
        Set(value As Integer)
            piAttempts = value
        End Set
    End Property
End Class
```

The following code defines a **VB.NET** class named `clsHWGrade` that inherits from another class named `clsGrade`. The `clsHWGrade` class represents a homework grade and includes information about the submission date and the score.

Here's a breakdown of the code:

- `Public Class clsHWGrade`: This line declares a class named `clsHWGrade` intended to represent a homework grade.

- `Inherits clsGrade`: This line indicates that the `clsHWGrade` class inherits from another class called `clsGrade`. This suggests that the `clsGrade` class likely contains properties or methods related to grading, and `clsHWGrade` extends it to include homework-specific behavior.

- `Private pdSubmitted As Date`: This line declares a private field named `pdSubmitted` of type `Date`, which stores the homework submission date.

- `Public Sub New(piScore As Integer, pdSubmitted As Date)`: This is a constructor for the `clsHWGrade` class. It takes two parameters: `piScore` and `pdSubmitted`. When an instance of `clsHWGrade` is created, this constructor is called to initialize its properties.

- `Me.Score = piScore`: This line sets the `Score` property of the `clsHWGrade` instance to the value of the `piScore` parameter. The `Score` property is inherited from the `clsGrade` class.

- `Me.pdSubmitted = pdSubmitted`: This line sets the `pdSubmitted` field of the `clsHWGrade` instance to the value of the `pdSubmitted` parameter.

- `Public Property Submitted As Date`: This line defines a public property named `Submitted`. This property allows getting and setting the submission date of the homework.

- `Get`: This block defines the getter for the `Submitted` property. When the property is accessed, this block is executed and returns the value of the `pdSubmitted` private field.

- `Set(value As Date)`: This block defines the setter for the `Submitted` property. When the property is assigned a value, this block is executed and assigns the provided value to the `pdSubmitted` private field.

This class is designed to work with homework grades and includes an additional attribute for the homework submission date. The inheritance from the `clsGrade` class provides common grading-related functionality, and `clsHWGrade` extends it to include homework-specific behavior. Here is the code:

```
Public Class clsHWGrade
    Inherits clsGrade
    Private pdSubmitted As Date

    Public Sub New(piScore As Integer, pdSubmitted As Date)
        Me.Score = piScore
        Me.pdSubmitted = pdSubmitted
    End Sub

    Public Property Submitted As Date
        Get
            Return pdSubmitted
        End Get
        Set(value As Date)
            pdSubmitted = value
        End Set
    End Property
End Class
```

The following code reads data from a CSV file containing student grades and calculates the average scores for each student's assignments (tests, quizzes, and homework). It uses dictionaries and lists to organize and process the data. Here's a breakdown of the code:

- `Imports System.IO` imports the `System.IO` namespace, which provides classes for working with input and output operations, including file reading

- `Module Module1` defines a VB.NET module named `Module1`

- `Sub Main()` is the program's entry point, where the execution starts

- `Dim filePath As String = "..\..\students.csv"` defines the relative path to the CSV file containing student grades

- `Dim studentGrades As New Dictionary(Of String, List(Of clsGrade))()` initializes a dictionary named `studentGrades` that associates student names (keys) with lists of grade objects (values)

- `Using reader As New StreamReader(filePath)` begins a block for reading the CSV file using `StreamReader`

- The `While Not reader.EndOfStream` loop iterates through each line in the CSV file

- `Dim line As String = reader.ReadLine()` reads a line from the CSV file

- `Dim fields As String() = line.Split(",")` splits the line into an array of strings using commas as delimiters

- `If fields.Length = 4 Then` checks whether four fields are in the line

- Inside the `If` block, the code extracts the student name, score, category, and additional information from the `fields` array

- Depending on the category, a new instance of `clsTestGrade`, `clsQuizGrade`, or `clsHWGrade` is created and added to the appropriate student's list in the `student Grades` dictionary

- The loop inside the `Using` block continues until all lines of the CSV file are processed

- After reading the CSV file, the code calculates and displays average scores for each student and category

- The outer `For Each` loop iterates through each student in the `studentGrades` dictionary

- Inside the loop, it calculates the total scores and counts for each category (`Test`, `Quiz`, and HW) using separate variables

- It then calculates the average scores for each category, handling cases where the count is zero

- Finally, it prints out the student's name and the calculated average scores for tests, quizzes, and homework

- `Console.ReadLine()` waits for the user to press *Enter* before exiting the program

In summary, this code reads student grade data from a CSV file, organizes it in a dictionary, calculates average scores for different assignment categories, and displays the results for each student:

```
Imports System.IO
```

```vbnet
Module Module1

    Sub Main()

        Dim filePath As String = "..\..\students.csv"
        Dim studentGrades As New Dictionary(Of String,
          List(Of Grade))()

        Using reader As New StreamReader(filePath)
            While Not reader.EndOfStream
                Dim line As String = reader.ReadLine()
                Dim fields As String() = line.Split(",")

                If fields.Length = 4 Then
                    Dim studentName As String =
                      fields(0).Trim()
                    If Not studentGrades.ContainsKey
                      (studentName) Then
                        studentGrades.Add(studentName, New
                          List(Of Grade)())
                    End If
                    Dim score As String = fields(1).Trim()
                    Dim category As String =
                      fields(2).Trim()
                    Select Case category
                        Case "Test"
                            studentGrades(studentName).
                              Add(New TestGrade(score,
                                fields(3).Trim()))
                        Case "Quiz"
                            studentGrades(studentName).
                              Add(New QuizGrade(score,
                                fields(3).Trim()))
                        Case "HW"
                            studentGrades(studentName).
                              Add(New HWGrade(score,
                                fields(3).Trim()))
                    End Select


                End If
            End While
        End Using
```

```
        For Each studentName As String In
          studentGrades.Keys
            Dim testtotal, testcnt, quiztotal, quizcnt,
              hwtotal, hwcnt As Integer
            Dim grades As List(Of Grade) = studentGrades
              (studentName)
            For Each eagrade As Grade In grades
                If TypeOf eagrade Is TestGrade Then
                    testtotal += eagrade.Score
                    testcnt += 1
                ElseIf TypeOf eagrade Is QuizGrade Then
                    quiztotal += eagrade.Score
                    quizcnt += 1
                Else
                    hwtotal += eagrade.Score
                    hwcnt += 1
                End If
            Next
            Dim testavg, quizavg, hwavg As Double
            If testcnt > 0 Then
                testavg = testtotal / testcnt
            End If
            If quizcnt > 0 Then
                quizavg = quiztotal / quizcnt
            End If
            If hwcnt > 0 Then
                hwavg = hwtotal / hwcnt
            End If

            Console.WriteLine(studentName & ": Test Avg:" &
              testavg.ToString("N1") &
                " Quiz Avg:" & quizavg.ToString("N1") &
                " HW Avg:" & hwavg.ToString("N1"))

        Next

        Console.ReadLine()
    End Sub

End Module
```

Next, let's plug this code into the IDE to give it a try.

Coding the Visual Basic.NET console grading tool

The following steps allow you to code and execute a VB.NET Windows Forms version of `Hello World`:

1.   Start **Microsoft Visual Studio**.

2.   Choose **Create New Project**.

3.   Choose **Console (.NET Framework)** and click **Next**.

4.   Accept the default project name and solution name.

5.   Change the location if you prefer your code in a different folder.

6.   Accept the default framework.

7.   Click **Create Project**.

8.   Enter the code example from the preceding section in the main module.

9.   Create class files for each class described earlier with the filename being the same as the name of the class with an extension of vb.

10. To run your program, click the green triangle on the toolbar above your code.

The next section will tackle the Object-Oriented Grade Tool in ASP.NET.

Building the project in ASP.NET

What follows is **ASP.NET VB.NET** code for the Object-Oriented Grading Tool. The code is very similar to the earlier code presented for the **VB.NET** console application. With this example, no ASPX code changes are required as the output is handled with the `response.write` method:

```
Imports System.IO

Public Class ASPNETOOGradingTool
    Inherits System.Web.UI.Page

    Protected Sub Page_Load(ByVal sender As Object, ByVal e
      As System.EventArgs) Handles Me.Load
        Dim filePath As String = "students.csv"
        Dim studentGrades As New Dictionary(Of String,
          List(Of clsGrade))()

        Using reader As New StreamReader(filePath)
            While Not reader.EndOfStream
                Dim line As String = reader.ReadLine()
                Dim fields As String() = line.Split(",")
```

```vb
            If fields.Length = 4 Then
                Dim studentName As String =
                    fields(0).Trim()
                If Not studentGrades.ContainsKey
                    (studentName) Then
                        studentGrades.Add(studentName, New
                            List(Of Grade)())
                End If
                Dim score As String = fields(1).Trim()
                Dim category As String =
                    fields(2).Trim()
                Select Case category
                    Case "Test"
                            studentGrades(studentName).
                                Add(New clsTestGrade(score,
                                    fields(3).Trim()))
                    Case "Quiz"
                            studentGrades(studentName).
                                Add(New clsQuizGrade(score,
                                    fields(3).Trim()))
                    Case "HW"
                            studentGrades(studentName).
                                Add(New clsHWGrade(score,
                                    fields(3).Trim()))
                End Select


            End If
        End While
End Using

For Each studentName As String In
    studentGrades.Keys
        Dim testtotal, testcnt, quiztotal, quizcnt,
            hwtotal, hwcnt As Integer
        Dim grades As List(Of clsGrade) =
            studentGrades(studentName)
        For Each eagrade As clsGrade In grades
            If TypeOf eagrade Is clsTestGrade Then
                testtotal += eagrade.Score
                testcnt += 1
            ElseIf TypeOf eagrade Is clsQuizGrade Then
                quiztotal += eagrade.Score
```

```
                    quizcnt += 1
            Else
                hwtotal += eagrade.Score
                hwcnt += 1
            End If
        Next
        Dim testavg, quizavg, hwavg As Double
        If testcnt > 0 Then
            testavg = testtotal / testcnt
        End If
        If quizcnt > 0 Then
            quizavg = quiztotal / quizcnt
        End If
        If hwcnt > 0 Then
            hwavg = hwtotal / hwcnt
        End If

        Response.Write(studentName & ": Test Avg:" &
          testavg.ToString("N1") &
            " Quiz Avg:" & quizavg.ToString("N1") &
            " HW Avg:" & hwavg.ToString("N1") &
              "</br>")


      Next

    End Sub

  End Class
```

Next, let's plug this code into Visual Studio.

Coding an ASP.NET calculator with Visual Studio

The following steps will allow you to code and execute an ASP.NET web page that displays `Hello World`:

1. Start **Microsoft Visual Studio 2019**.
2. Choose **Create New Project**.
3. Choose **ASP.NET Web Application** and click **Next**.
4. Accept the default project name and solution name.
5. Change the location if you prefer your code in a different folder.
6. Accept the default framework.

7. Click **Create Project**.

8. Choose **Web Forms**.

9. When the designer loads, navigate to the solution explorer and right-click on the solution.

10. Choose **Add Item** from the pop-up menu.

11. Choose **Web Form Visual Basic**.

12. Right-click on the ASPX source and choose **View Code** from the pop-up menu.

13. Enter the VB.NET code from the preceding code example that falls in the `Page_load` procedure.

14. Create class files for each class described, ensuring the filename is the same as the name of the class with an extension of `vb`.

15. To run your program, click the green triangle on the toolbar above your code.

Summary

In this chapter, we utilized the skills we have gained to build an Object-Oriented Grading Tool for the newer Visual Basic family members that support object inheritance. This project used skills from *Project II* in *Chapter 11* and added the use of inheritance and polymorphism. The next chapter will look at the Request and Response model used in modern web application architectures.

Part 4:
Server-Side Development

In the fourth part of this book, we will learn to use Visual Basic in web development. Server-side development focuses on coding in response to HTTP requests from a web browser. We will learn to format an answer in an HTTP response. We will conclude the book in this part with a summary of everything we have learned, along with some projects and learning you can do in the future.

This part has the following chapters:

17

The Request and Response Model

In **Classic ASP** and **ASP.NET**, the Request and Response model describes how client-server communication is handled. Both Classic ASP and ASP.NET are web application frameworks that allow you to build dynamic web applications using server-side technologies. The Request and Response model is at the core of this communication process.

The following is an explanation of the Request part of the model:

- The request represents the information sent by the client (usually a web browser) to the server. It includes **HTTP** headers, form data, query parameters, cookies, and more.

- When a user accesses a web page or sends a form submission, the client generates an HTTP request and sends it to the server.

- The server-side code in your application processes this request and responds accordingly.

- Classic ASP and ASP.NET provide various objects and properties to access the request data.

The following is an explanation of the Response part of the model:

- The response represents the information sent by the server back to the client in response to the client's request

- The server-side code in your Classic ASP and ASP.NET applications generates the appropriate response, which is then returned to the client

- The response typically includes **HTML** content, **CSS** styles, **JavaScript**, and other resources necessary for rendering a web page

- Classic ASP and ASP.NET provide objects and methods to manage the response

Classic ASP and ASP.NET use a combination of `Request` and `Response` objects to process client requests, execute server-side logic, and generate dynamic HTML or other content to be sent back to the client. This model enables the creation of interactive and dynamic web applications.

In this chapter, we're going to cover the following main topics:

- Classic ASP `Request` object
- ASP.NET `Request` object
- Classic ASP `Response` object
- ASP.NET `Response` object

Classic ASP Request object

In Classic ASP, the `Request` object is a built-in server-side object that's used to retrieve information sent by the client (usually a web browser) to the web server. The `Request` object allows you to access various data types, including form data, query parameters, cookies, and more, making it essential for processing client requests and building dynamic web applications.

Here are some common properties and methods of the `Request` object in Classic ASP:

- **Form collection** (`Request.Form`): Provides access to the data sent to the server using the **HTTP POST** method (usually from HTML forms).

  Example: `Request.Form("username")` returns the value submitted with the input field named `"username"`.

- **Query string** (`Request.QueryString`): Provides access to the data sent to the server using the **HTTP GET** method (usually from URLs).

  Example: `Request.QueryString("id")` returns the value of the `"id"` parameter from the URL.

- **Cookies collection** (`Request.Cookies`): Provides access to the client's cookies sent with the request.

  Example: `Request.Cookies("username").Value` returns the value of the `"username"` cookie.

- **ServerVariables collection** (`Request.ServerVariables`): Provides access to server-related information, such as server name, server port, client IP address, and more.

  Example: `Request.ServerVariables("REMOTE_ADDR")` returns the client's IP address.

- **HTTP method** (`Request.ServerVariables("REQUEST_METHOD")`): Returns the HTTP method used in the request (for example, `GET` or `POST`).

- **Path** (`Request.ServerVariables("PATH_INFO")`): Returns the virtual path of the requested ASP file. When used for web development, this information comes from the URL.

- **Query string** (`Request.ServerVariables("QUERY_STRING")`): Returns the raw query string from the URL.

You don't need to create the `Request` object in Classic ASP to use it explicitly; it's automatically available in ASP pages. You can access its properties and methods directly within your ASP code.

Here's a simple example that demonstrates the use of the `Request` object in Classic ASP:

```
<%
Dim username
username = Request.Form("username")

If Len(username) > 0 Then
    Response.Write "Hello, " & username & "!"
Else
%>
    <form method="post">
        <label for="username">Enter your name:</label>
        <input type="text" name="username" id="username">
        <input type="submit" value="Submit">
    </form>
<%
End If
%>
```

In this example, when the user submits the form with their name, `Request.Form("username")` retrieves the value from the form data, and the personalized greeting is displayed using `Response.Write`. If no name is provided, the form is displayed for the user to input their name.

Next, we will examine the **ASP.NET** `Request` object.

ASP.NET Request object

In ASP.NET, the `Request` object is a server-side object that's used to retrieve information sent by the client (usually a web browser) to the web server. It provides access to various data types, including form data, query parameters, cookies, headers, and so on. The `Request` object is essential to processing client requests and building dynamic web applications in ASP.NET.

The `Request` object is accessible in all ASP.NET page types, including ASPX pages, code-behind files, and custom handlers. It is available by default, and you can use it directly without creating an instance.

Here are some common properties and methods of the `Request` object in ASP.NET. They are very close to the Classic ASP properties and methods:

- **Form collection** (`Request.Form`): Provides access to the data sent to the server using the **HTTP POST** method (usually from HTML forms).

Example: `Request.Form["username"]` returns the value submitted with the input field named `"username"`.

- **Query string** (`Request.QueryString`): Provides access to the data sent to the server using the **HTTP GET** method (usually from URLs).

  Example: `Request.QueryString["id"]` returns the value of the `"id"` parameter from the URL.

- **Cookies collection** (`Request.Cookies`): Provides access to the client's cookies sent with the request.

  Example: `Request.Cookies["username"].Value` returns the value of the `"username"` cookie.

- **Headers collection** (`Request.Headers`): Provides access to the HTTP headers sent by the client.

  Example: `Request.Headers["User-Agent"]` returns the `User-Agent` header of the client's browser.

- **HTTP method** (`Request.HttpMethod`): Returns the HTTP method used in the request (for example, GET, POST).

- **Path** (`Request.Path`): Returns the virtual path of the requested ASP.NET page.

- **QueryString** (`Request.QueryString.ToString()`): Returns the raw query string from the URL.

- **InputStream** (`Request.InputStream`): Provides access to the raw request body as a stream, which helps handle non-form data requests.

To use the `Request` object in ASP.NET, you can directly access its properties and methods within your ASPX pages or code-behind files.

Here's a simple example that demonstrates the use of the `Request` object in an ASP.NET page:

```
<%@ Page Language="VB" AutoEventWireup="false" CodeBehind="Default.
aspx.vb" Inherits="WebApplication.Default" %>

<!DOCTYPE html>
<html>
<head>
    <title>ASP.NET Request Object Example</title>
</head>
<body>
    <form id="form1" runat="server">
        <div>
```

```
        <asp:Label ID="lblMessage"
            runat="server"></asp:Label>
        <asp:TextBox ID="txtUsername"
            runat="server"></asp:TextBox>
        <asp:Button ID="btnSubmit" runat="server"
            Text="Submit" OnClick="btnSubmit_Click" />
      </div>
    </form>
  </body>
</html>
```

Here's the **VB.NET** code-behind (`Default.aspx.vb`):

```
Imports System

Namespace WebApplication
    Public Class _Default
        Inherits System.Web.UI.Page

        Protected Sub btnSubmit_Click(ByVal sender As
          Object, ByVal e As EventArgs)
            Dim username As String = Request.Form
              ("txtUsername")
            If Not String.IsNullOrEmpty(username) Then
                lblMessage.Text = "Hello, " & username &
                  "!"
            End If
        End Sub
    End Class
End Namespace
```

In the previous example, the ASPX page contains a simple form with a text box and a submit button. When the user enters their name in the text box and clicks the **Submit** button, the `btnSubmit_Click` event handler in the VB.NET code-behind is executed – `Request.Form("txtUsername")` retrieves the value from the form data (note that we are using the `txtUsername` control ID to access the value). The personalized greeting is displayed on the web page using `lblMessage` if a name is provided.

The `Request` object is automatically available in the VB.NET code-behind file, and you can access its properties (for example, `Request.Form`, `Request.Cookies`, and so on) to handle client-side data and build dynamic ASP.NET applications.

Next, we will examine the **Classic ASP** `Response` object.

Classic ASP Response object

In Classic ASP, the Response object is a server-side object that's used to send output from the server back to the client (usually a web browser) as a response to the client's request. The Response object provides methods and properties to set HTTP headers, send content (such as HTML, text, or binary data), manage cookies, and perform various other operations related to generating the response.

The Response object is automatically available in Classic ASP and does not require creating an instance. You can use it directly within your ASP code to customize the response that's sent to the client.

Here are some standard methods and properties of the Response object in Classic ASP:

- Write and WriteText: The Write method sends output to the client, such as HTML content, text, or variable values. The WriteText method sends the client binary data (such as images).

- Redirect: The Redirect method redirects the client to a different URL.

- AddHeader: The AddHeader method adds custom HTTP headers to the response.

- ContentType: The ContentType property sets the MIME type of the response content.

- Cookies collection (Response.Cookies): This manages cookies that are sent back to the client.

- Clear: The Clear method clears any buffered HTML content before sending the response.

- BinaryWrite: The BinaryWrite method sends binary data directly to the client, such as images or files.

Here's a simple example that demonstrates the use of the Response object in Classic ASP:

```
<%
Dim username
username = Request.Form("username")

If Len(username) > 0 Then
    Response.Write "Hello, " & username & "!"
Else
%>
    <form method="post">
        <label for="username">Enter your name:</label>
        <input type="text" name="username" id="username">
        <input type="submit" value="Submit">
    </form>
<%
End If
%>
```

In the previous example, when the user submits a form with their name, the `Response.Write` method sends back a personalized greeting to the client. If no name is provided, an HTML form is sent back to the user so that they can input their name.

The `Response` object is an essential tool for controlling the output that's sent to the client in Classic ASP. The `Response` object enables developers to build dynamic and interactive web applications.

Next, we will examine the **ASP.NET** `Response` object.

ASP.NET Response object

In ASP.NET, the `Response` object is a server-side object that's used to send output from the server back to the client (usually a web browser) as a response to the client's request. The `Response` object provides methods and properties to set HTTP headers, send content (such as HTML, text, or binary data), manage cookies, and perform various other operations related to generating the response.

The `Response` object is automatically available in all ASP.NET page types, including ASPX pages, code-behind files, and custom handlers. You can use it directly within your ASP.NET code to customize the response that's sent to the client.

Here are some standard methods and properties of the `Response` object in ASP.NET. Again, these are very similar to the Classic ASP `Response` methods and properties:

- `Write`: The `Write` method, as in the Classic ASP `Response` object, sends output to the client, such as HTML content, text, or variable values

- `WriteFile`: The `WriteFile` method sends binary data, such as images or files, directly to the client

- `Redirect`: The `Redirect` method redirects the client to a different URL

- `AddHeader`: The `AddHeader` method adds custom HTTP headers to the response

- `ContentType`: The `ContentType` property is used to set the MIME type of the response content

- `Cookies collection` (`Response.Cookies`): This manages cookies that are sent back to the client

- `Clear`: The `Clear` method clears any buffered HTML content before sending the response

- `BinaryWrite`: The `BinaryWrite` method sends binary data, such as images or files, directly to the client

- `Output caching`: The `Response.Cache` property allows you to configure caching options for the response, enabling better performance by serving cached content to subsequent requests

Here's a simple example that demonstrates the use of the `Response` object in an **ASP.NET** page.

ASPX page (`Default.aspx`):

```
<%@ Page Language="VB" AutoEventWireup="true" CodeBehind="Default.
aspx.vb" Inherits="WebApplication.Default" %>

<!DOCTYPE html>
<html>
<head>
    <title>ASP.NET Response Object Example</title>
</head>
<body>
    <form id="form1" runat="server">
        <div>
            <asp:Button ID="btnGreet" runat="server"
              Text="Greet" OnClick="btnGreet_Click" />
        </div>
    </form>
</body>
</html>
```

VB.NET code-behind (`Default.aspx.vb`):

```
Imports System

Namespace WebApplication
    Public Class _Default
        Inherits System.Web.UI.Page

        Protected Sub btnGreet_Click(ByVal sender As
          Object, ByVal e As EventArgs)
            Dim username As String = "John" ' You can get
              the username from user input or any other
                source
            Response.Write("Hello, " & username & "!")
        End Sub
    End Class
End Namespace
```

In this example, when the user clicks the **Greet** button, the `btnGreet_Click` event handler in the VB.NET code-behind is executed. The `Response.Write` method sends a personalized greeting to the client, displaying `"Hello, John!"` on the web page.

The `Response` object in ASP.NET is a useful tool for controlling the output that's sent to the client, allowing developers to build dynamic and interactive web applications from the code-behind files.

Summary

This chapter introduced the Request and Response model that's used in programming web applications in VB. We looked at the methods and properties for both Classic ASP and ASP.NET. Most new development today follows this Request and Response model, so it will be very important in your programming future. The next chapter will look at variable scope in web applications developed with VB.

18

Variable Scope and Concurrency

Like many other server-side web technologies, **Active Server Pages (ASP)** follows the **Hypertext Transfer Protocol (HTTP)** stateless model. **HTTP** is the foundation of data communication on the World Wide Web. It is a stateless protocol, which means that each request made by a client to the server is independent and unrelated to previous requests. The server does not retain any information about the client's previous requests, and each request is processed in isolation.

Statelessness in HTTP has some important implications:

- **No client context**: The server does not keep track of the client's state between requests. Each request contains all the necessary information for the server to process it.

- **Scalability**: Stateless protocols are more scalable because servers do not need to maintain client state, allowing them to handle many concurrent requests without memory overhead.

- **No inherent session management**: Since no session management exists, developers must implement mechanisms such as cookies or tokens to maintain the session state across multiple requests.

To maintain user state and provide a more interactive experience, web applications often use techniques such as cookies, URL parameters, or hidden form fields to store session data on the client side. When a client sends a request to the server, this session data is included, allowing the server to identify and associate the request with the correct session.

Concerning **ASP**, it can leverage various methods to manage session state in a stateless HTTP environment, including the following:

- **Cookies**: ASP can use cookies to store small amounts of data on the client side, which can be sent back to the server with each request

- **Session objects**: ASP provides a session object, which allows developers to store and retrieve data per session, typically utilizing server-side memory or a designated session store

- **Query strings or hidden form fields**: Data can be returned to the server through URLs or hidden form fields within **HTML** forms

These mechanisms allow ASP applications to simulate statefulness within the stateless HTTP model, providing users with a more dynamic and personalized experience.

In this chapter, we're going to cover the following main topics:

- Session scope
- Application scope
- Request scope and cookies

Session scope

In ASP, session scope refers to the duration data is maintained and accessible for a particular user session. A web development session is the period between a user's initial request to a web application and their final interaction or until the session times out due to inactivity.

ASP provides a `session` object, allowing developers to store and retrieve data specific to a user session. This data can be accessed throughout the session, allowing for stateful behavior within the stateless HTTP protocol. The session data, typically in memory, is stored on the server side and is associated with a unique session identifier (often stored in a cookie on the client side) that enables the server to identify the user's session for subsequent requests.

Here's a basic overview of how session scope works in ASP:

- **Session start**: When a user accesses an ASP application for the first time or starts a new session, the server assigns a unique session ID to the user. This session ID is stored in a cookie on the client's browser or passed through URL parameters in subsequent requests.

- **Storing data**: During the user's interaction with the application, you can use the `Session` object in ASP to store data specific to that session. For example, you can store user preferences, shopping cart items, authentication details, or any other information relevant to the user's session:

```
<%
    ' Storing data in session scope
    Session("username") = "JohnDoe"
    Session("cart_items") = Array("item1", "item2",
      "item3")
%>
```

- **Accessing data**: The stored session data can be accessed and used throughout the user's session. This allows you to maintain state and personalize the user experience based on their previous interactions:

```
<%
    ' Accessing data from session scope
    Dim username
    username = Session("username")
    Response.Write("Welcome, " & username)
%>
```

- **End of session**: When the user ends their session (either by logging out, closing the browser, or after a certain period of inactivity), the session data is cleared, and the session is terminated. The server releases the resources associated with that session.

It's essential to use session scope judiciously and avoid storing large amounts of data in session objects as it can lead to increased memory usage on the server and potentially impact application performance. Instead, it is advisable to use session scope for essential user-specific data that needs to persist across multiple requests during a session.

Application scope

In ASP, application scope refers to a shared data storage area that persists throughout the lifetime of an ASP application. Unlike session scope, which stores data specific to each user session, application scope allows data to be shared among all users accessing the application. The data stored in the application scope is available to all users, regardless of their sessions.

The application scope is used when you must maintain consistent global or shared data across all user sessions and requests. Everyday use cases for the application scope include storing configuration settings, shared resources, or any data that should be accessed and updated globally across the entire application.

Here's a basic overview of how the application scope works in ASP:

- **Data initialization**: When the ASP application starts, you can use the `Application` object to initialize and store data in the application scope. This is typically done in the `global.asa` file or in the `Application_OnStart` event handler:

```
<!-- Global.asa -->
<SCRIPT LANGUAGE="VBScript" RUNAT="Server">
Sub Application_OnStart
    ' Initialize data in the Application scope
    Application("SiteName") = "My ASP Website"
    Application("VisitorsCount") = 0
End Sub
</SCRIPT>
```

- **Accessing data**: Once data is stored in the application scope, it can be accessed and used by any user accessing the application. For example, you can display the site's name and the total number of visitors across the complete application:

```
<%
    ' Accessing data from the Application scope
    Dim siteName
    siteName = Application("SiteName")
    Response.Write("Welcome to " & siteName & "<br>")

    Dim visitorsCount
    visitorsCount = Application("VisitorsCount")
    Response.Write("Total visitors: " & visitorsCount)
%>
```

- **Shared updates**: Since the application scope is shared among all users, you must be careful when updating data. If multiple users try to modify the same application data simultaneously, it can lead to conflicts and unexpected results. To ensure data integrity, you should implement synchronization mechanisms such as `Application.Lock` and `Application.Unlock` to prevent concurrent access:

```
<%
    ' Increment visitors count in a thread-safe manner
    Application.Lock
    Application("VisitorsCount") =
        Application("VisitorsCount") + 1
    Application.Unlock
%>
```

- **Application shutdown**: When the ASP application is shut down or restarted, the data in the application scope is cleared. This can happen when the application pool is recycled or the server is restarted.

It's essential to use the application scope judiciously and only store data that needs to be shared globally across the entire application. Be cautious not to overload the application scope with excessive data as it can increase memory usage and impact the application's performance.

Request scope and cookies

In ASP, the request scope and cookies are two essential concepts for web development and handling user interactions. Let's explore both in more detail.

Request scope in ASP

The request scope refers to the lifetime and accessibility of data during a single HTTP request-response cycle in ASP. When a client (typically a web browser) sends a request to the server, the server processes the request and generates a response. During this process, data can be passed from the client to the server (for example, form data, query strings, and HTTP headers) and vice versa (for example, server-generated content).

ASP provides an intrinsic `Request` object that allows developers to access data coming from the client in the HTTP request. This object provides various properties and methods to access request data, such as form data, query strings, cookies, and HTTP headers.

The following is an example of accessing form data in ASP using the `Request` object:

```
<%
    Dim username
    username = Request.Form("username")
    Response.Write("Hello, " & username)
%>
```

Next, let's look at an example of accessing query string data in ASP using the `Request` object:

URL (dummy): *http://127.0.0.1/page.asp?name=Aspen*:

```
<%
    Dim name
    name = Request.QueryString("name")
    Response.Write("Hello, " & name)
%>
```

The request scope is limited to a single request-response cycle. Once the response is sent to the client, the data in the request scope is discarded.

Cookies in ASP

Cookies are small pieces of data that are stored on the client side (usually within the client's web browser) and are sent back to the server with each HTTP request. They allow the server to remember stateful information about the client between different requests, even in a stateless HTTP protocol.

ASP provides a way to work with cookies using the `Response` and `Request` objects. The `Response` object is used to set or modify cookies, while the `Request` object is used to retrieve and access cookies sent by the client.

The following is an example of setting a cookie in ASP:

```
<%
    ' Set a cookie that expires in 30 days
    Response.Cookies("username") = "AspenOlmsted"
    Response.Cookies("username").Expires = DateAdd("d", 30,
       Now())
%>
```

The following example deals with retrieving a cookie in ASP:

```
<%
    Dim username
    username = Request.Cookies("username")
    If len(username) > 0 Then
        Response.Write("Welcome back, " & username)
    Else
        Response.Write("Welcome, guest!")
    End If
%>
```

Cookies are commonly used for various purposes, such as maintaining user sessions, storing user preferences, tracking user behavior, and implementing personalized website experiences.

It's essential to use cookies judiciously and be mindful of privacy concerns. Some users may disable or block cookies in their browsers, and it's essential to handle scenarios where cookies are not available gracefully. Additionally, sensitive information should not be stored in cookies as they are visible to the client and can be manipulated by users. For sensitive data, I prefer using server-side session storage mechanisms.

Summary

This chapter introduced variable scope and concurrency while programming web applications in **VB**. We looked at defining the application, session, and request scopes for variable lifetimes in the web application. We also discussed cookies, which we can use to pass small amounts of data between the client and the server. The next chapter, *Chapter 19*, will utilize these new skills in our final programming project.

19

Project Part IV

In our fourth project, we will use the skills we have learned about the HTTP request-response model, as well as variable scope and concurrency, to build a server-side grade averaging tool that will receive grades one at a time from the client. In addition, it will send back the average for all the grades it received.

In this chapter, we're going to cover the following main topics:

- Developing a server-side grade averaging tool in Classic ASP
- Developing a server-side grade averaging tool in ASP.NET

Technical requirements

Please complete the steps for installing the code with the VB family member from *Chapter 1*. All the example code for this chapter is available in the following GitHub repository: https://github. com/PacktPublishing/Learn-Visual-Basics-Quick-Start-Guide-/tree/ main/ProjectPartIV.

Building the project in Classic ASP

The following is a **Classic ASP** program that accepts grades from the client web browser and responds with the average of the grades to this point. The program will accept grades in a web browser and send them to the server, which will calculate a new average and return it:

```
<FORM>
<INPUT name="grade"/>
<INPUT type="submit" value="Add Grade"/>
</FORM>
<%
if Request("grade").Count > 0 Then
    Session("Total") = Session("Total") +
        CInt(Request("grade"))
    Session("Cnt") = Session("Cnt") + 1
```

```
      response.write("Grade Average: ")
      response.write(Session("Total") / Session("Cnt"))
  End If
  %>
```

The first part of the code defines an **HTML** form that allows users to input a grade value. It contains a text input field named grade and a submit button labeled Add Grade.

The next section of the code uses Classic ASP to process the form submission and calculate the average grade.

The if statement checks whether any data has been submitted with the name grade using the Request("grade").Count condition. If data is present, it means the form was submitted.

Inside the if block, we have the following:

- Session("Total") is a session variable that keeps track of the sum of grades. It adds the value of the submitted grade (Request("grade")) to the current value.

- Session("Cnt") is another session variable that keeps track of the number of grades. It increments by one to count the number of grades entered.

- response.write is used to output text to the web page. In this case, it displays Grade Average, followed by the calculated average, which is the total sum divided by the count.

In summary, this code creates a simple web form that allows users to input grades. When the form is submitted, the Classic ASP code calculates and displays the average grade by maintaining a running sum of grades and a count of grades entered using session variables.

Coding the Classic ASP grading tool with Notepad

Follow these steps to code and execute VBScript that displays the calculator developed in this project:

1. Start Notepad.
2. Enter all the code from the code example in the previous section.
3. Save the file as ClassicASPSSGrades.asp in the root web directory (this is typically c:\ inetpub\wwwroot).
4. Open a web browser.
5. Enter http://127.0.0.1/ClassicASPSSGrades.asp.

The next section will tackle the server-side grade average tool example in **ASP.NET**.

Building the project in ASP.NET

What follows is **ASP.NET** and **VB.NET** code for the server-side grade averaging tool. The first piece of code is the ASPX code:

```
<%@ Page Language="vb" AutoEventWireup="false"
CodeBehind="ASPNETSSGrades.aspx.vb " Inherits="WebApplication8.
WebForm1" %>

<!DOCTYPE html>

<html xmlns="http://www.w3.org/1999/xhtml">
<head runat="server">
    <title></title>
</head>
<body>
    <form id="form1" runat="server">
        <div>
            <asp:TextBox ID="tbGrade"
              runat="server"></asp:TextBox>
            <asp:Button ID="Button1" runat="server"
              Text="Add Grade" />
        </div>
        <asp:Label ID="lblAverage"
          runat="server"></asp:Label>
    </form>
</body>
</html>
```

The first line of the preceding code block is a directive at the top of the page that provides instructions to the **ASP.NET** runtime. It specifies the page's language (**Visual Basic**), turns off the automatic event wire-up (we'll discuss this later), and points to the code-behind file (`ASPNETSSGrades.aspx.vb`).

The next section of code contains the **HTML** markup for the web page.

The page starts with the usual HTML structure and includes a `<form>` element with an ID of `form1` that runs on the server.

Inside the form, there's a `<div>` element containing two ASP.NET controls:

- An `<asp:TextBox>` control with an ID of `tbGrade`. This is a text input where the user can enter a grade.
- An `<asp:Button>` control with an ID of `Button1` and a piece of text stating `Add Grade`. This is a button that users can click to perform an action (such as adding a grade).

Below the `<div>` element, there's an `<asp:Label>` control with an ID of `lblAverage`. This is where the calculated average grade will be displayed.

The `runat="server"` attribute on these HTML elements indicates that they are ASP.NET server controls, which means they can be manipulated and interacted with via the server-side code.

The following is the code from the code-behind file, `ASPNETSSGrades.aspx.vb`:

```
Public Class WebForm1
    Inherits System.Web.UI.Page

    Protected Sub Page_Load(ByVal sender As Object, ByVal e
      As System.EventArgs) Handles Me.Load

    End Sub

    Protected Sub Button1_Click(sender As Object, e As
      EventArgs) Handles Button1.Click
        Session("Total") += CInt(tbGrade.Text)
        Session("Cnt") += 1
        lblAverage.Text = "Grade Average: " &
          Session("Total") / Session("Cnt")
    End Sub
End Class
```

The beginning of the previous code declares a class named `WebForm1` that inherits from the `System.Web.UI.Page` class. This indicates that `WebForm1` is an ASP.NET page and can utilize the properties and methods provided by the base class.

The section of code that starts with `Protected Sub Page_Load` is an event handler for the page's Load event. The Load event is raised when the page is being loaded. This event is often used for initializing page-related data or controls. In this case, the event handler is empty, so it doesn't contain any specific code to execute when the page loads.

The section of code that starts with `Protected Sub Button1_Click` is an event handler for the `Click` event of a control with an ID of `Button1` (as specified in the `Handles Button1.Click` part). This event is triggered when the user clicks the button. Here's what the code inside the event handler does:

- It uses the `CInt` function to convert the text entered in a control with an ID of `tbGrade` to an integer and adds it to the value stored in the `Total` session variable

- It increments the value stored in the `Cnt` session variable by 1

- It calculates the average grade by dividing the total sum stored in the `Total` session variable by the count stored in the `Cnt` session variable

- The calculated average is then assigned to the `Text` property of a control with an ID of `lblAverage`, which will display the calculated average on the web page

The code uses session variables to store and maintain the running total and count of grades across different requests and interactions. The calculated average is dynamically displayed on the web page when the user clicks the Button1 button.

Coding the ASP.NET server-side grade averager with Visual Studio

Follow these steps to code and execute an ASP.NET web page that displays Hello World:

1. Start **Microsoft Visual Studio 2019**.
2. Choose **Create New Project**.
3. Choose **ASP.NET Web Application** and click **Next**.
4. Accept **Default Project Name** and **Solution Name**.
5. Change the location if you want your code to be in a different folder.
6. Accept the default **Framework** option.
7. Click **Create Project**.
8. Choose **Web Forms**.
9. When the designer loads, navigate to **Solution Explorer** and right-click on the solution.
10. Choose **Add Item** from the pop-up menu.
11. Choose **Web Form Visual Basic**.
12. Choose **Source**.
13. Paste in the ASPX code within the **FORM** tag.
14. Right-click on the ASPX source and choose **View Code** from the pop-up menu.
15. Enter the VB.NET code example from the previous section that falls in the Button1_ click procedure.
16. To run your program, click the green triangle on the toolbar above your code.

Try to internalize the similarities between the Classic ASP and ASP.NET versions of the code. As you can see, ASP.NET separates the HTML from VB to try to make maintenance easier. This becomes very important as the project's size grows.

Summary

In this chapter, we utilized the skills we gained in the previous chapters regarding the HTTP request-response model, variable scope, and concurrency to build a server-side student grade averaging tool for both **Classic ASP** and **ASP.NET**. In the next chapter, we will conclude this book by summarizing what we have covered, thinking about other VB topics you can study, and projects you can develop to demonstrate the skills you've learned throughout this book.

20
Conclusions

We have covered many different **Visual Basic (VB)** topics in this book. I want to close out this book by summarizing what we have learned while suggesting other areas of interest for you to study around VB and some suggestions for projects you could do to show a potential employer your new skills.

In this chapter, we're going to cover the following main topics:

- Summing it all up
- Further reading topics
- Suggested projects

Summing it all up

Throughout this book, we've looked at the use of VB in many different environments, including Windows Desktop development with Classic VB and VB.NET Console. We also looked at using VB as a server-side programming language in **Classic ASP** and **ASP.NET**. **Visual Basic for Applications (VBA)** was also used to embed VB in Microsoft Office products, and we explored using VB in scripts with **VBScript**.

In this section, we'll provide a brief review of the major topics we covered in this book.

In VB, variable declaration, type, scope, and assignment are fundamental concepts that are used to work with data and manage program logic. Here's an overview of each of these concepts.

Variable declaration

In VB, you declare variables using the `Dim` (short for **Dimension**) keyword. Variable declaration reserves memory space for storing data and associates a name with it. The basic syntax for declaring a variable is as follows:

```
Dim variableName As DataType
```

Data types

VB supports various data types to represent different kinds of values. Some common data types include the following:

- `Integer`: Represents whole numbers
- `Double` or `Single`: Represents floating-point numbers
- `String`: Represents text
- `Boolean`: Represents true or false values
- `Date`: Represents dates and times
- `Object`: Represents an instance of a class

Scope

The scope of a variable refers to where in the code the variable is accessible and can be used. There are different levels of scope in VB:

- **Procedure-level scope**: Only variables declared within a procedure (function or subroutine) are accessible.
- **Module-level scope**: Variables declared at the module level (outside of any procedure) are accessible to all procedures within that module.
- **Global scope**: Variables declared in a module with the `Public` keyword are accessible from any module or form in the project.
- **Assignment**: Assignment involves storing a value in a variable. You can assign values to variables using the assignment operator (=). Here's an example:

  ```
  age = 25
  name = "John"
  ```

- **Iteration**: Iteration, also known as looping, is a fundamental programming concept that allows you to execute a block of code repeatedly. In VB, you can use various constructs to implement iteration. Here are some standard methods for iteration:

  - The `For...Next` loop: The `For...Next` loop is used when you know the number of iterations in advance. It iterates through a range of values defined by a starting, ending, and optional step value:

    ```
    For counter As Integer = startValue To endValue Step
       stepValue
    Next
    ```

- The `While` loop: The `While` loop repeats a code block, so long as a specified condition remains true:

```
While condition
    ' Code to be executed in each iteration
End While
```

- The `Do...Loop` loop: The Do...Loop loop is similar to the `While` loop, but the condition is checked at the end. This means the loop's code block is always executed at least once:

```
Do
    ' Code to be executed in each iteration
Loop While condition
```

- The `For Each` loop: The `For Each` loop is used to iterate through elements in a collection, such as an array or a collection class:

```
For Each element As ElementType In collection
    ' Code to be executed in each iteration using
        'element'
Next
```

Decision branching

Decision branching, also known as conditional statements, is a core programming concept that allows you to control the flow of your program based on certain conditions. In VB, you can implement decision branching using the following conditional statements:

- The `If...Then` statement: The `If...Then` statement allows you to execute a code block if a specified condition is true. You can also include an optional `Else` block to handle cases when the condition is false:

```
If condition Then
    ' Code to be executed if condition is true
Else
    ' Code to be executed if condition is false
End If
```

- The `If...Then...ElseIf...Else` statement: This statement allows you to test multiple conditions in sequence using `ElseIf` blocks. The code block associated with the first true condition is executed:

```
If condition1 Then
    ' Code to be executed if condition1 is true
ElseIf condition2 Then
    ' Code to be executed if condition2 is true
```

```
    Else
        ' Code to be executed if no condition is true
    End If
```

- The `Select Case` statement: The `Select Case` statement allows you to compare a single expression against multiple possible values. It provides a more concise way to handle numerous conditions:

```
Select Case expression
    Case value1
        ' Code to be executed if the expression
            matches value1
    Case value2
        ' Code to be executed if the expression
            matches value2
    Case Else
        ' Code to be executed if no match is found
End Select
```

Procedure modularization

In VB, procedure modularization involves breaking down your code into smaller, manageable pieces called procedures or functions. This practice helps improve code readability, reusability, and maintainability. There are two main types of procedures – `Sub` procedures (subs) and `Function` procedures (functions):

- `Sub`: Sub procedures, often called subs, are used to group a set of statements to perform a specific task. They don't return a value:

```
Sub ProcedureName(parameters)
    ' Code to perform a specific task
End Sub
```

- `Function`: Function procedures, or functions, are similar to subs but return a value. You can use them when you need to calculate or derive a result:

```
Function FunctionName(parameters) As DataType
    ' Code to calculate and return a value
End Function
```

Collections

In VB, data structures allow you to store and manage multiple elements of the same or different data types. Collections provide a convenient way to work with data groups and offer various methods for adding, removing, and accessing elements. VB provides several built-in collection classes that you can use to accomplish different tasks. We will not summarize the collection types here as they differ by VB family member, but you should revisit *Chapter 10* for a recap.

File handling

File handling in VB involves reading from and writing to files on your computer's filesystem. Each VB family member provides various classes and methods for handling files, allowing you to create, open, read, and write files. You can revisit *Chapter 9* for a recap.

Object encapsulation

Object encapsulation, also known as encapsulation or information hiding, is a fundamental principle of **object-oriented programming** (**OOP**) that emphasizes bundling data (attributes) and the methods (functions or procedures) that operate on that data into a single unit called an object. Encapsulation helps control access to an object's internal state and promotes modularization and abstraction. Please revisit *Chapter 12* to learn how to encapsulate variables in the different VB family members.

Inheritance

Inheritance is a fundamental concept in OOP that allows you to create new classes (called subclasses or derived classes) based on existing classes (called base classes or parent classes). Inheritance enables you to reuse and extend the functionality of existing classes, promoting code reusability and hierarchy. In VB, you can implement inheritance using the `Inherits` keyword or interface inheritance with interfaces. Please revisit *Chapter 13* to *Chapter 15* to see how inheritance is done in the different VB family members.

Polymorphism

Polymorphism is a core concept in OOP that allows objects of different classes to be treated as objects of a standard base class. It enables you to write more generic and flexible code that can work uniformly with various objects. Polymorphism is achieved through method overriding and method overloading.

In VB, polymorphism is primarily achieved through method overriding and interface implementation. Please revisit *Chapter 14* to learn how to use polymorphism in the different VB family members.

Web application HTTP model and concurrency

In web applications, communication between a client's web browser and a server occurs using the **Hypertext Transfer Protocol** (**HTTP**). This model involves the client sending requests to the server and the server responding with data. Please revisit *Chapter 17* and *Chapter 18* to learn how to use VB in Classic ASP and ASP.NET applications.

Next, let's look at some potentially interesting VB topics that were outside the scope of this book.

Further reading topics

In this book, we covered the core VB programming topics that are frequently used, but you may need to modify code that uses other technologies. I recommend that you familiarize yourself with the following topics, which we did not cover.

VB6 form development

VB6 was widely used for developing Windows applications, particularly **graphical user interface (GUI)** applications using forms. VB6 form development involved creating user interfaces by designing forms visually and adding code to implement the functionality. Here's a general overview of the process for VB6 form development:

- **Create a new project**: Open the VB6 IDE and create a new project. You can choose from various project templates, including Standard EXE, to create standalone applications.

- **Design forms**: VB6 provides a visual form designer where you can create your application's user interfaces. You can drag and drop various controls (buttons, textboxes, labels, and so on) onto the form, arrange them, and set their properties using the **Properties** window.

- **Add controls**: To add controls to a form, you can drag them from the toolbox onto the form. After adding controls, you can adjust their properties, such as size, location, text, font, and more, using the **Properties** window.

- **Event handling**: VB6 forms have various events (for example, `Click`, `Load`, and `Resize`) associated with them. You can double-click on a control or form to open the code window and automatically generate an event handler for that control or form. You can also write the code that will be executed when the event occurs in the code window. For instance, clicking a button could trigger code to perform a specific action.

- **Debugging**: The VB6 IDE includes debugging tools that allow you to set breakpoints, step through code, inspect variables, and identify and fix errors in your code.

- **Build and deploy**: Once your application is ready, you can make it into an executable file. VB6 provides options for compiling and creating standalone executable files that could be run on Windows machines. You might also need to package additional files if your application uses external components or libraries.

- **Distribution**: Distributing VB6 applications requires packaging runtime libraries and components along with the executable to ensure the application will run on other machines without VB6 installed.

VB.NET WinForms development

VB.NET WinForms provides a framework for creating GUI applications using forms and controls. Here's an overview of the process:

- **Create a new project**: Open Visual Studio and create a new **Windows Forms Application** project using **VB.NET**.

- **Design forms**: VB.NET WinForms still uses a visual form designer. You can drag and drop controls from the toolbox onto the form, adjust their properties, and arrange them as needed.

- **Add controls**: Add controls such as buttons, textboxes, labels, and more by dragging them onto the form. You can then set their properties in the **Properties** window.

- **Event handling**: Similar to VB6, VB.NET WinForms allows you to double-click controls to generate event handler code automatically. You can write code for events such as `Click`, `Load`, `Resize`, and more to control the behavior of your application.

- **Debugging**: Visual Studio offers powerful debugging tools for VB.NET WinForms applications. You can set breakpoints, step through code, inspect variables, and use various debugging windows to identify and fix issues.

- **Data binding**: VB.NET WinForms applications often use data binding to connect controls such as `DataGridViews` or `ListBoxes` to data sources. This simplifies tasks such as displaying and editing data from databases or collections.

- **Build and deploy**: You can build your VB.NET WinForms application into an executable file or other deployable format. Visual Studio provides options to generate installer packages that include runtime components.

- **Distribution**: When distributing VB.NET WinForms applications, you must ensure that the appropriate version of the **.NET** Framework is installed on users' machines. Modern versions of Windows typically include this framework by default.

COM object creation

Creating **Component Object Model (COM)** objects in VB.NET involves creating classes that adhere to the COM specifications, allowing them to be used by other languages and applications that support COM, such as VB6 and VBA. COM objects can be used in a wide range of scenarios, from extending the functionality of existing software to creating reusable components. Here's an overview of the process of creating a COM object in VB.NET:

- **Create a new Class Library project**: Open Visual Studio and create a new **Class Library** project using VB.NET.

- **Define the COM class**: Define a class you want to expose as a COM object. This class should be marked with attributes that indicate it's a COM class. The most essential attributes are [ClassInterface], [Guid], and [ComVisible]:

```
Imports System.Runtime.InteropServices

<ClassInterface(ClassInterfaceType.AutoDual)>
<Guid("YOUR-GUID-HERE")>
<ComVisible(True)>
Public Class MyCOMClass
    ' Define properties, methods, and events here.
End Class
```

Replace "YOUR-GUID-HERE" with an actual GUID for your class. You can generate a GUID using tools such as Visual Studio's **Create GUID** utility.

- **Define properties, methods, and events**: Inside your COM class, define properties, methods, and events that you want to expose to COM clients.

- **Build the project**: Build the project to create the class library assembly.

- **Register the assembly**: To use the COM object, you need to register the assembly using the Windows Regasm.exe command-line tool. Open the Command Prompt as an administrator and run the following command:

```
Regasm.exe /codebase YourAssembly.dll
```

Replace YourAssembly.dll with the actual name of your assembly.

- **Use the COM object**: Once registered, the COM object can be used by other applications that support COM. You can create instances of your COM class in applications such as VB6 and C#, or even scripting languages such as VBScript. Here is an example of its usage in VBScript:

```
Dim obj
Set obj = CreateObject("YourNamespace.YourCOMClass")
obj.YourMethod()
Set obj = Nothing
```

Next, let's look at some projects you could tackle in VB to show your new skills to potential employers.

Suggested projects

I recommend that you continue your learning by tackling a larger project you can use to demonstrate your new skills to potential employers. The most important thing to your success is that you are motivated by the creative project. Try to pick an idea that is something useful to you. Here are a few ideas:

- Try to develop a command-line tool that processes files. File processing is a very important automation for both pleasure and business. For example, thousands of volunteers modify games on older consoles to use newer data. An example of this is a sports game that has new rosters updated yearly. You could try to develop a program to automate this. Be sure to take it slowly and not try to do too much. It is best if the process can be done manually by updating the files and you are just making a utility to do the manual steps in code.

- Try to develop a more complicated web application than the simple samples we developed in this book. There are many terrific books on HTML that you can pick up to build up your knowledge. This could be an interactive game such as a board game or a utility that assists you in managing your life.

- Try to develop a startup program that you can launch when Microsoft Windows starts. If there is an activity you do on your machine daily, such as backing up or cleaning up files, see if you can do this in code.

These are just a few ideas to motivate your creativity. The important thing is that you use the new mind muscles you have gained on projects you care about.

Summary

This chapter concludes our look at VB and the different family members Microsoft has created over the years. VB has had a huge impact on the way we use and program computers today. I want to thank you for working hard to learn the material presented in this book. Keep up the great work in developing your technical skills!

Index

www.packtpub.com

Subscribe to our online digital library for full access to over 7,000 books and videos, as well as industry leading tools to help you plan your personal development and advance your career. For more information, please visit our website.

Why subscribe?

- Spend less time learning and more time coding with practical eBooks and Videos from over 4,000 industry professionals

- Improve your learning with Skill Plans built especially for you

- Get a free eBook or video every month

- Fully searchable for easy access to vital information

- Copy and paste, print, and bookmark content

Did you know that Packt offers eBook versions of every book published, with PDF and ePub files available? You can upgrade to the eBook version at www.packtpub.com and as a print book customer, you are entitled to a discount on the eBook copy. Get in touch with us at customercare@packtpub.com for more details.

At www.packtpub.com, you can also read a collection of free technical articles, sign up for a range of free newsletters, and receive exclusive discounts and offers on Packt books and eBooks.

Other Books You May Enjoy

If you enjoyed this book, you may be interested in these other books by Packt:

.NET MAUI Cross-Platform Application Development

Roger Ye

ISBN: 978-1-80056-922-5

- Discover the latest features of .NET 6 that can be used in mobile and desktop app development
- Find out how to build cross-platform apps with .NET MAUI and Blazor
- Implement device-specific features using .NET MAUI Essentials
- Integrate third-party libraries and add your own device-specific features
- Discover .NET class unit test using xUnit.net and Razor components unit test using bUnit
- Deploy apps in different app stores on mobile as well as desktop

.NET MAUI for C# Developers

Jesse Liberty, Rodrigo Juarez

ISBN: 978-1-83763-169-8

- Explore the fundamentals of creating .NET MAUI apps with Visual Studio
- Understand XAML as the key tool for building your user interface
- Obtain and display data using layout and controls
- Discover the MVVM pattern to create robust apps
- Acquire the skills for storing and retrieving persistent data
- Use unit testing to ensure your app is solid and reliable

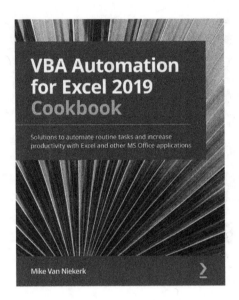

VBA Automation for Excel 2019 Cookbook

Mike Van Niekerk

ISBN: 978-1-78961-003-1

- Understand the VBA programming language's role in the context of the MS Office suite
- Discover various aspects of VBA programming such as its terminology, syntax, procedures, functions, and forms
- Investigate the elements, features, and characteristics of the VBA Editor to write and edit custom scripts
- Automate Excel sheets with the help of ranges
- Explore error handling and debugging techniques to catch bugs in your programs
- Create and use custom dialog boxes to collect data from users
- Customize and extend Office apps such as Excel, PowerPoint, and Word

Packt is searching for authors like you

If you're interested in becoming an author for Packt, please visit `authors.packtpub.com` and apply today. We have worked with thousands of developers and tech professionals, just like you, to help them share their insight with the global tech community. You can make a general application, apply for a specific hot topic that we are recruiting an author for, or submit your own idea.

Share your thoughts

Now you've finished *Visual Basic Quickstart Guide*, we'd love to hear your thoughts! Scan the QR code below to go straight to the Amazon review page for this book and share your feedback or leave a review on the site that you purchased it from.

`https://packt.link/r/1805125311`

Your review is important to us and the tech community and will help us make sure we're delivering excellent quality content.

Download a free PDF copy of this book

Thanks for purchasing this book!

Do you like to read on the go but are unable to carry your print books everywhere?

Is your eBook purchase not compatible with the device of your choice?

Don't worry, now with every Packt book you get a DRM-free PDF version of that book at no cost.

Read anywhere, any place, on any device. Search, copy, and paste code from your favorite technical books directly into your application.

The perks don't stop there, you can get exclusive access to discounts, newsletters, and great free content in your inbox daily

Follow these simple steps to get the benefits:

1. Scan the QR code or visit the link below

https://packt.link/free-ebook/9781805125310

2. Submit your proof of purchase
3. That's it! We'll send your free PDF and other benefits to your email directly